INTERNATIONAL SERIES OF MONOGRAPHS ON
PURE AND APPLIED BIOLOGY

Division: **MODERN TRENDS IN PHYSIOLOGICAL SCIENCES**

GENERAL EDITORS: P. ALEXANDER and Z. M. BACQ

VOLUME 17

A

OTHER TITLES IN THE MODERN TRENDS IN PHYSIOLOGICAL SCIENCES DIVISION

General Editors: P. ALEXANDER and Z. M. BACQ

OTHER DIVISIONS IN THE SERIES ON PURE AND APPLIED BIOLOGY

BIOCHEMISTRY

BOTANY

PLANT PHYSIOLOGY

ZOOLOGY

CHEMICAL PROTECTION OF THE BODY
AGAINST IONIZING RADIATION

CHEMICAL PROTECTION OF THE BODY AGAINST IONIZING RADIATION

Edited by

V. S. BALABUKHA

Translated from the Russian by

J. T. GREAVES

Translation edited by

J. H. BARNES

MEMBER OF SCIENTIFIC STAFF
M.R.C., RADIOBIOLOGICAL RESEARCH UNIT
HARWELL

A Pergamon Press Book

THE MACMILLAN COMPANY
NEW YORK
1963

THE MACMILLAN COMPANY
60 Fifth Avenue
New York 11, N.Y.

This book is distributed by
THE MACMILLAN COMPANY
pursuant to a special arrangement with
PERGAMON PRESS
Oxford · England

Library of Congress Card Number 63–10104

This book is a translation of the original Russian
Khimicheskaya zashchita organizma ot ioniziruyushchikh izluchenii
(Published in Moscow, 1960, by Atomizdat)

PRINTED IN GREAT BRITAIN BY
CHORLEY & PICKERSGILL LIMITED, LEEDS

Contents

Preface

THE FIRST part of this collection of papers is devoted to the problem of chemical protection from penetrating radiations. In this section there is a short summary of the present status of the problem. In the experimental work data are given on the synthesis and biological testing of the protective properties of a number of chemical compounds (aminothiols and pyrimidine derivatives). Attention is paid mainly to an explanation of the mechanism of action of the protective substances.

A study of this mechanism is of substantial value both for an understanding of the nature of radiation damage to living material, and for a choice of effective chemical agents for the prevention of radiation disease. In the papers, possible ways of protecting the tissues from the harmful action of ionizing radiation are discussed in the light of some radiobiological and biophysical hypotheses.

The second part of the collection contains experimental investigations dealing with the problem of the removal of radioactive isotopes from the body. The characteristics of the state of certain radioactive isotopes in the blood and bone tissue are given and on a basis of physico-chemical results and biological experiments, the effectiveness of complexing agents as preparations instrumental in combining with and removing these isotopes from the body is estimated.

The collection is intended for chemists working in research on chemical protective agents and complexing agents, and also for biologists and other specialists occupied with problems of radiobiology.

Preface to the English Edition

THE PROBLEM of the chemical protection of the body from the action of radiation is at the present stage a very important section of radiobiology and attracts the attention of a wide circle of research workers.

Problems connected with the elimination of ingested radioactive isotopes from the body are no less important and are of great practical significance.

The authors agreed that it would be expedient to publish some experimental results on this problem.

In Part I it is proposed that the attention of readers of the book be drawn to the investigations on laboratory animals and to the level of radioprotective effectiveness of a number of chemical compounds.

Possible mechanisms of the action of the chemical protectors are judged on a basis of the results obtained.

In Part II problems of eliminating radioactive isotopes from the body are discussed.

Experimental material characterizing the nature of the binding of certain isotopes in the blood and bone tissue is given. The effect of complex-forming agents on this binding is explained.

The effectiveness of a number of complexones is estimated from this aspect.

Since this collection was written, new investigations on the given problems have appeared, and have been submitted for discussion at international conferences, symposia and in individual published papers (investigations of Weiss, Dale, Grey, Alexander, Langendorff, and Dutch, Norwegian and Soviet radiobiologists), in which different points of view on the mechanism of the action of chemical protectors have been reflected.

On the other hand, considerable supplementary material on the testing of new, more powerful complex-forming agents has been accumulated (investigations of Catsch, Schubert and co-workers, and of Soviet authors).

Nevertheless, the material cited in the book may also be of interest to specialists in radiobiology and may prove useful for clearing up problems of the chemical protection of the body against ionizing radiation.

V. S. BALABUKHA

PART I

CHEMICAL PROTECTION
AGAINST IONIZING RADIATIONS

Present State of Chemical Protection against Penetrating Radiation

V. S. BALABUKHA

IN CONNECTION with the increasing use of atomic energy in the national economy, questions of protection, mainly chemical, from penetrating radiations in particular, are acquiring very great importance.

During the last decade intensified searches have been carried out for chemical compounds, the administration of which before exposure may protect the body from the harmful effects of radiations.

There is already wide literature devoted to the synthesis and choice of chemical preparations having such protective properties in experiments on laboratory animals. Investigations on the mechanism of action of protective substances are being extended, which is of great significance mainly insofar as the protective effect is directly connected with one or other action of the protective substances on the fundamental radiochemical processes which develop in the tissues under the influence of ionizing radiation. It is not by chance that the protective substances as a rule exert a favourable effect on the course of radiation sickness if they are introduced before exposure, but are not effective if they are introduced after exposure.

The role of the protective substances evidently amounts to a prevention of the secondary reactions which are the results of the primary radiochemical processes induced by exposure. If the nature of the primary biological effect of radiation in such a complex system as the living body or even a single cell is still not fully disclosed, the mechanism of action of chemical prophylactics is even less completely known. At the present time there does not exist a single universal theory explaining the intimate mechanism of the prophylactic action of effective chemical compounds. A number of hypotheses have been put forward, to a certain extent substantiated, attempting to explain the action of the protective substances from biophysical, biochemical and pharmacological points of view. However, in the interpretations of

3

experimental results there remain some inherent contradictions which do not permit these hypotheses to be united into a single whole.

In a short introductory article it is not necessary to give an exhaustive review of all the investigations dealing with the nature of chemical prophylaxis in radiation sickness.

The reader can turn to a number of fairly full reviews and monographs on this problem, which have appeared in recent years (e.g. Refs. 4, 5, 7, 9, 15–18, etc.). In what follows only certain results of work in this direction are mentioned.

The numerous theories concerning the possible mechanism of the action of protective substances, expounded by different research workers and to a certain extent confirmed experimentally, amount to a few basic propositions on which it is necessary to dwell briefly.

1.   It is an established fact, based on a whole series of observations, that the oxygen effect is present during irradiation. In the initial processes of the radiation syndrome the main part is played by oxidation reactions induced by oxidizing radicals which arise in the aqueous phases of the body under the influence of exposure, and a number of authors[3, 15, 19, 21] think that the protective effect is directly connected with a temporary reduction in the partial pressure of oxygen in the cells and tissues of the body. The reduction in partial pressure is achieved by introducing certain prophylactics, i.e. by causing in the body as a whole or in its individual organs a condition of hypoxia which decreases the possibility of the formation of oxidizing radicals.

2.   Other research workers[5] consider that the role of prophylactics amounts to capture of the oxidizing radicals; on reacting with them, the prophylactics protect the tissues from the harmful effect of the radicals. This point of view has recently been developed further.

It is suggested that the oxidizing reactions lead to the formation of organic peroxides and hydroperoxides of a lipoid character, with the formation finally of products which have a toxic effect[4]. In studying the antioxidative properties of organic compounds used as prophylactics, certain authors come to the conclusion that these compounds prevent the development of subsequent oxidizing chain-reactions[2].

3.   In the literature the idea is most frequently put forward that the prophylactics (aminothiol series) are capable of "masking" the tissue sites which are sensitive to the action of radiation. Entering into temporary combination with the thiol groups of proteins and enzymes

they form mixed disulphides, thus protecting the biologically important molecules from the harmful effect of the oxidizing radicals or hydroperoxides (Refs. 12, 13, etc.).

It has also been proposed that prophylactics can react at the double bonds of unsaturated aliphatic acids, pyrimidine bases, and so on, thus protecting molecules sensitive to radiation (see present symposium, p. 51).

This possible mechanism of the action of prophylactic chemicals is at present attracting the continuous attention of research workers and demands correspondingly wide experimental confirmation.

4. An idea is developing that prophylactic action may be related to the possibility of the transfer of energy absorbed by a protein molecule along the carbon chain. The temporary disulphide links (tissue site —S— S-prophylactic), formed during the action of thiol prophylactics, have the capacity of taking up absorbed energy. Moreover, rupture of the S—S-link takes place with splitting off of the prophylactic and its subsequent oxidation, as a result of which the radiation energy is eliminated[20].

According to the expression of Eldjarn, Pihl and Shapiro[13], "non-destructive" dissipation of energy in the ionized protein molecule takes place.

5. Finally, yet another hypothesis has recently been developed concerning the possibility of absorbing the energy of secondary electromagnetic radiations (ultraviolet emission), which arise as a result of the exchange of energy.

Chemical compounds which have maximum absorption in the ultraviolet region of the spectrum can, on account of this faculty, absorb quanta of ultraviolet radiations characterized by high chemical activity. Similar substances, which thus appear to be "traps" for ultraviolet radiation, can protect the nucleic acids from the development of harmful photochemical reactions in them. As is known, the maximum of absorption of nucleic acids lies just in this region of the spectrum (see present symposium p. 98).

Such in general are the present-day ideas concerning the possible mechanism of the action of chemical prophylactics. These ideas no doubt need further experimental checking, and the investigations of recent years in the chemical prophylactic field have been directed mainly towards this.

Among the numerous chemical compounds studied in relation to their prophylactic properties against ionizing radiation, those which are of the greatest interest and practical significance at the present time are the low molecular-weight aminothiol compounds. They have free or unstably bound (potential) SH- and NH_2-groups (cysteine and its derivatives, aminothiols of the β-mercapto-ethylamine (MEA) type, isothiouronium compounds and some others). For prophylactics of this series the general formula

$$\begin{matrix} R^1 \\ \diagdown \\ \diagup \\ R^2 \end{matrix} N—(CH_2)_x—SH$$

is suggested, where $x \gtreqless 3$ and R^1 and R^2 are hydrogen or alkyl residues. SH-α-amino acids also have prophylactic action.

It has been found that substitution for hydrogen in the amino and especially the SH-group leads to a reduction in prophylactic properties. It has been shown in particular that aromatic substituents in the amino group (R^1 and R^2 being phenyl and benzyl residues) abolish the protective action of these compounds completely. An increase in the length of the carbon chain leads to a drop in prophylactic activity[16, 17].

The specificity of structure of the molecule as a whole for the realization of protection may be emphasized and a number of examples are given where a small change in the structure of the molecule alters the properties of such a compound as a prophylactic.

For example, if cysteine

$$\left(HS—\overset{\beta}{CH_2}\overset{\alpha}{CH}—COOH \atop \qquad\quad | \atop \qquad\;\; NH_2 \right)$$

and α-homocysteine

$$\left(HS—\overset{\gamma}{CH_2}—\overset{\beta}{CH_2}—\overset{\alpha}{CH}—COOH \atop \qquad\qquad\qquad\quad | \atop \qquad\qquad\qquad NH_2 \right)$$

have prophylactic properties, then isocysteine

$$\left(H_2N—\overset{\beta}{CH_2}—\overset{\alpha}{CH}—COOH \atop \qquad\qquad\;\; | \atop \qquad\qquad\; SH \right)$$

and β-homocysteine

$$\left(\text{HS}-\overset{\gamma}{\text{C}}\text{H}_2-\overset{\beta}{\underset{\underset{\text{NH}_2}{|}}{\text{C}}}\text{H}-\overset{\alpha}{\text{C}}\text{H}_2-\text{COOH} \right)$$

are compounds which sensitize animals to the action of radiation. The same is observed in the case of cysteamine $(\text{HS}-\text{CH}_2-\text{CH}_2-\text{NH}_2)$ and thioglycol $(\text{HS}-\text{CH}_2-\text{CH}_2-\text{OH})$.

As regards the effect of the SH-group on the prophylactic action, it should be noted that there is no direct parallel between the number of SH-groups in a thiol compound and its prophylactic efficiency, but a definite correlation exists between the accumulation of SH-compounds in the organs and the prophylactic effect. V. G. Yakovlev and L. S. Isupova (see present symposium p. 39) demonstrated this clearly. The rate of penetration of the thiol compound into the tissues of the organs, where the primary processes of protection develop, is of substantial significance. V. G. Yakovlev's experiments (see present symposium p. 11) confirm this position graphically.

Isupova's investigations showed, both by means of labelled l-cysteine and by direct determination of SH-groups in the organs, that amino-thiol prophylactics penetrate into the testes in smaller amounts than into liver tissue. Similar observations are given in the literature for labelled MEA[11].

It is characteristic that compounds containing a free SH-group, but not having an amino-group in their composition, do not possess prophylactic action. An example of this is thioglycollic acid, the administration of which leads to an accumulation of SH-groups in the liver tissue without the manifestation of any prophylactic effect (see p. 49). This once again confirms the specificity of the "aminothiol" structure of prophylactics. Evidently just this structure is necessary for the formation of temporary "protective" compounds with the tissue sites, including mixed disulphides (see p. 71). Data exist regarding the formation of disulphides, protein—S—S-prophylactic, in experiments *in vitro* and *in vivo* (in the blood)[11a, 11].

In the literature great attention is paid to the problem of the discrepancy between the total amount of SH-groups in the tissues and the quantity of thiol groups introduced in the form of the aminothiol compounds for a protective purpose.

Investigators explain this discrepancy by suggesting that not all the SH-groups of the tissues are equally sensitive to exposure, and they are not all capable of entering with the same ease into reaction with the thiol groups of the prophylactics. This problem still requires very intensive study.

The irregular distribution of prophylactics in the body (local accumulation of prophylactic and protection of one organ or another) may evidently be of great significance.

If prophylactics are approached from the point of view of their effect on the consumption of oxygen by the animal body, then the range of these compounds is widening.

The reduction in oxygen consumption is brought about in different ways. For prophylactic purposes it is possible to introduce certain methaemoglobin-forming agents (sodium nitrite, p-aminopropiophenone, acetanilide), enzyme poisons which inhibit respiration of the tissue (cyanides, malononitrile, sodium azide) or biologically active amines, including cysteamine, which exert an influence on the respiratory centres.

These compounds reduce the consumption of oxygen by the body[3], and in experiments mainly on small animals they have a certain prophylactic effect. Apparently, only aminothiol compounds will be of practical significance.

Nevertheless, this problem as a whole still requires detailed investigations of a more precise nature. For example, there are indications that the formation of methaemoglobin under the influence of certain chemical protectors (sodium nitrite) and the prophylactic effect are not coincident in time[8].

Ye. F. Romantsev showed that the condition of reduced oxygen consumption after the administration of a number of prophylactics persists for a fairly long period (6 hr or more), whereas their prophylactic action is limited to a shorter interval of time before the moment of exposure.

There are hardly any investigations which establish the existence of tissue hypoxia caused by the administration of prophylactics. Nevertheless, attention has recently been paid to the substantial uniformity of the prophylactic action of these compounds, which has a definite connection with the oxygen effect. There are reports of the influence of aminothiols on the formation of organic peroxides of lipoid character in the body, initiated by the action of ionizing radiation. Ye. F. Romantsev and Z. I. Zhulanova (see present symposium p. 86) showed that β-mercaptoethylamine reduces the formation of these peroxide compounds in normal animals and particularly in exposed animals. Evidently aminothiols, as antioxidants, prevent the development of oxidative chain-reactions induced by oxidizing radicals[4], which was clearly shown by Zhuravlev[2].

A new trend in work on chemical prophylaxis is the search for chemical compounds which can be used as "traps" for quanta of secondary ultraviolet radiations.

This trend, presented in G. Ye. Fradkin's paper, opens new possibilities in the selection of effective chemical protectors and widens the range of prophylactics by including compounds of the pyrimidine series.

As regards pharmacological modes of action of various chemical compounds used for purposes of protection, and their effects on the metabolic processes of the body as a whole, special studies have been devoted to these problems[1, 5, 6, 9, 10, 14].

On the basis of literature data and the experiments described in the papers of this symposium it may evidently be concluded that the mechanism of the protective action of chemical compounds has many aspects.

Besides the purely pharmacological effects of prophylactics and their effects on metabolic processes and on primary radiochemical reactions, it is also necessary to consider the possibility of the absorption of energy of secondary radiations. The question of what processes have the greatest importance in the prophylaxis of radiation sickness is a subject for further study.

In the discussion of experimental results the biochemical, radiobiological and biophysical points of view are given.

These papers are of a discursive character, but they should attract the attention of experimenters working in the field of chemical protection.

REFERENCES

1. ARBUZOV, S. YA., BARYSHNIKOV, I. I., GENERALOV, V. I. and MUKHIN, YE. A., *Theses of reports of the All-Union Conf. on Med. Radiology* (Tezisy dokladov Vsesoyuznoi konferentsii po med. radiologii), 25 (1956).

B

2. ZHURAVLEV, A. I., *Theses of reports on the scientific conference on the problem: Pathogenesis, clinical aspects, therapy and prophylaxis of radiation disease* (Tezisy dokladov nauchnoi konferentsii po probleme: Patogenez, klinika, terapiya i profilaktika luchevoi bolezni), 35 (1957).

3. ROMANTSEV, YE. F. and SAVICH, A. V., *Chemical Protection from the Action of Ionizing Radiation* (Khimicheskaya ashchita ot deistviya ioniziruyushchei radiatsii), Moscow, Medgiz (1958).

4. TARUSOV, B. N., *Primary Processes of Radiation Disease* (Pervichnye protsessy luchevogo porazheniya), Moscow, Medgiz (1957).

5. BACQ, Z. M. and ALEXANDER, P., *Fundamentals of Radiobiology*, Butterworths, London (1955).

6. BACQ, Z. M. and FISCHER, P., *Arch. Intern. Physiol.*, **61**, 3, 417 (1953).

7. BOND, V. P. and CRONKITE, E. P., *Ann. Rev. Physiol.*, **19**, 299 (1957).

8. COLE, L. J. and ELLIS, M. E., *Amer. J. Physiol.*, **175**, 3, 429 (1953).

9. DELLA BELLA, D. and BACQ, Z. M., *Arch. Exper. Path. Pharmakol.*, **219**, 366 (1953).

10. DELLA BELLA, D., GOFFART, M. and BACQ, Z. M., *Arch. Intern. Physiol.*, **61**, 3, 449 (1953).

11. ELDJARN, L. and PIHL, A., *J. Biol. Chem.*, **223**, 1, 341 (1954).

11a. ELDJARN, L. and PIHL, A., *J. Biol. Chem.*, **225**, 3, 499 (1957).

12. ELDJARN, L. and NYGAARD, O., *Arch. Intern. Physiol.*, **62**, 4, 476 (1956).

13. ELDJARN, L., PIHL, A. and SHAPIRO, B., *Proc. Intern. Conf. Peaceful Uses of Atomic Energy*, Geneva (1955).

14. GOFFART, M. and DELLA BELLA, D., *Arch. Intern. Physiol.*, **62**, 4, 455 (1954).

15. GRAY, Z. H., Progress in Radiobiology, *Proc. Fourth Conf. on Radiobiol.*, Oliver & Boyd, p. 267 (1956).

16. HAGEN, U. and KOCH, R., *Z. Naturf.*, B.12b, **4**, 240 (1957).

17. KOCH, R., *Forsch. f. Geb. Rontgenstrahlen u. Nukl. Medizin*, **85**, 6, 767 (1956).

18. PATT, H. M., *Physiol. Rev.*, **33**, 2 (1953).

19. PATT, H. M., *Ann. Rev. Physiol.*, **16**, 51 (1954).

20. PIHL, A. and ELDJARN, L., *Intern. Conf. Radiobiol.*, Stockholm (1956).

21. SALERNO, P. R., UYEKI, E. and FRIEDELL, H. L., *Radiat. Res.*, **3**, 344 (1955).

Relationship between the Structure and Properties of Certain Compounds containing Sulphur and their Prophylactic Action against Penetrating Radiations

V. G. Yakovlev

OF THE organic compounds investigated as means for the protection of animals from the action of penetrating radiation from outside, substances containing non-oxidized sulphur are being studied intensively. Among them the most effective have been low molecular-weight compounds having free thiol (SH-) groups or those capable of liberating these groups in living tissues as a result of chemical reaction with the tissues. However, far from all SH-compounds have prophylactic action *in vivo*. An important condition for increasing the radio-resistance of the body is a definite "protective" structure of the thiol agent and the presence in the molecule of additional functional groups: —NH₂, —CO—NH— and so on. In Table 1 chemical compounds are listed, some with and some without prophylactic effect.

In the SH-compounds which have pronounced prophylactic action, it is present in very unequal degrees. It depends on even small changes in composition and structure of the molecules and on changes, closely connected with these, in chemical and physico-chemical properties and in biological action of the prophylactics. Therefore a study of the relationship between the degree of prophylactic effect and the properties of prophylactics may be of value in clarifying the biological mechanisms of chemical protection[3, 11, 12, 31, 41] from penetrating radiations. In connection with this, investigations were carried out in two directions:

1. Testing the prophylactic action of a number of compounds containing sulphur (mainly sulphydryl) on animals. For this purpose derivatives of thiazolidine and mercaptopropionic acid, substances of the semithioacetal type, salts of esters of cysteine with alcohols and certain other compounds were synthesized and tested.

TABLE 1

COMPARISON OF EFFECTIVE (PROTECTIVE) AND INEFFECTIVE
CHEMICAL COMPOUNDS

Effective compounds	Literature reference	Ineffective compounds	Literature reference
Cysteine: l; d; d,l (?)	Own data and also (20, 25, 28, 29, 35, 36, 42, 43, 45, 50–52)	d,l-Serine (hydroxyl analogue of cysteine)	Own data
Salts of esters of l-cysteine with aliphatic alcohols C_1—C_5	Own data	d,l-Isocysteine	Own data
α-Homocysteine	(31, 32, 35)	β-Methyl-; β,β-dimethyl-cysteine	(32)
Salts of β-mercapto-ethylamine (cysteamine)	Own data and also (4, 6, 27, 34, 45)	N,S-Alkyl(-acyl) cysteamines Ethanolamines	(35) Own data
α-Methyl-, β-Methyl-derivatives of β-mercaptoethylamine	(14)	Aminothiols with a carbon chain of more than 3 C atoms	(31)
Salts of disulphides of esters of cysteine and of the simplest aminothiols	Own data and also (2, 3, 6, 27)	SH-Pantetheine Thioglycollic acid d,l-Methionine and its esters with alcohols	(35) Own data Own data
S,β-Aminoethyliso-thiouronium Br. HBr	Own data and also (23, 30, 49)	Dimethyl-mercapto-pyrimidine	(24, 37)
N-β-Alanylcysteamine	(35)	Ergothioneine Thiolhistidine	(3) (37)

2. Comparison of the efficiency of prophylactics with their properties — solubility, ability to penetrate and capacity for forming complexes, and also the effect on the degree of accumulation and dynamics of non-protein SH-groups in radiosensitive organs and tissues.

The syntheses were carried out by methods described in the literature. As a result of the investigation, modifications were indicated which were used in the preparation of substances not known in the chemical literature. The compounds were identified by certain constants (melting points, solubility in water, etc.). For the unknown substances, quantitative analysis was carried out with determination of the total sulphur.

Experiments on chemical protection were carried out on white rats, male and female, weighing 200–230 g, maintained on an ordinary diet. "Protected" rats, and also rats which had received as "standard prophylactic" the same preparation of purified *l*-cysteine, were exposed in identical conditions with the control animals to filtered X-rays in doses of 650 r, close to the minimum 100 per cent lethal dose (the mortality in the control animals was 97–100 per cent). The exposure was carried out on a RUM-3 apparatus in the following conditions: voltage 180 kV, current strength 15 mA, filter 0·5 mm Cu + 1 mm Al, focal length 40 cm, dose-rate 34–35 r/min. The substances to be investigated were introduced in different doses intraperitoneally and orally at different times before exposure in the form of solutions in sterile water (pH 6·8–7·0) calculated at 1 ml/100 g body weight. The prophylactic effect was judged from the survival rate 30–40 days after exposure, and also from the change in weight of the surviving animals and the average length of life of the rats that had died from radiation sickness.

Non-protein sulphydryl groups were determined by the method of MacDonnel et al[38]. The following modifications were introduced into the method: (a) to prevent catalytic oxidation of the SH-groups by the oxygen of the air the proteinless filtrates of tissues during preparation, and also the solution for back titration, were stabilized by the addition of ethylenediaminetetraacetic acid (it does not interfere with the determination); (b) *l*-cysteine ethyl ester hydrochloride was taken as "standard" SH-compound for the back titration. This substance is characterized by very high constancy of composition, can easily be purified by recrystallization and is stable on keeping. Details of the method are reported in the present symposium (p. 55).

In the investigation of derivatives of mercaptopropionic acid, tests were carried out with different doses of the substances and methods of introducing them (in each experiment 8–12 rats were used). A small prophylactic effect was noted only for thiolactic acid. Keto- and hydroxy-derivatives (Table 2) have no prophylactic effect, although mercaptopyruvic acid is evidently one of the normal intermediate products of the metabolism of cysteine in the body of mammals[39].

TABLE 2

RESULTS OF A STUDY OF THE PROTECTIVE ACTION OF DERIVATIVES OF MERCAPTOPROPIONIC ACID

Acid	Chemical structure	Reference to method of synthesis	Protective action
d,l-α-Mercaptopropionic (thiolactic)	CH_3—CH—COOH | SH	(7)	20–30%
β-Mercaptopropionic (Thiohydracrylic)	CH_2—CH_2—COOH | SH	(10)	Not effective
d,l-α-Hydroxy-β-mercapto-propionic (thioglyceric)	CH_2—CH—COOH | | SH OH	(33)	Not effective
α-Keto-β-mercaptopropionic (mercaptopyruvic)	CH_2—C—COOH | || SH O	(40, 46)	—

During a more detailed study of thiolactic acid it was noted that its prophylactic effect changed somewhat, depending on whether this acid was tested in the free form or in the form of its salts. It is very probable that the efficiency of prophylactics is closely bound up with their ability to penetrate and breadth of distribution in the tissues and cells. Eardman[19] notes this, suggesting there is a difference between free glycollic acid and its sodium salt in ability to penetrate the envelope of the nuclei of eggs of fresh-water molluscs. As a result, apparently, of this the sodium salt had a greater prophylactic effect than the free acid. It is known from other literature sources[1] that the presence of a free carboxyl group reduces the ability of organic compounds to penetrate cell membranes and nuclear envelopes.

To study the problem of the role of penetration in the prophylactic effect of SH-compounds, we approached the problem by different methods. The properties of some salts of thiolcarboxylic acids were studied. For this purpose a prophylactic well known from radio-biological literature was used — reduced glutathione[13, 15, 16], which is a SH-tripeptide containing a carboxyl group capable of salt formation. In Table 3 the results of tests on the ammonium, sodium, potassium, lithium and magnesium salts of glutathione are given.

TABLE 3
RESULTS OF A STUDY OF THE PROTECTIVE ACTION OF DIFFERENT
SALTS OF SH-GLUTATHIONE

Protective compound	Injected into rat, mg/100 g weight			Survival rate, %
	—SH	—NH$_2$	metal	
SH-Glutathione–Na salt	27	13	18·7	20
SH-Glutathione–K salt	27	13	31·7	50
SH-Glutathione–NH$_4$ salt	16	7·8	—	50
SH-Glutathione–Li salt	16	7·8	3·4	70
Li acetate	—	—	3·4	0
SH-Glutathione–Mg salt	13	6·5	4·9	35

Note— Twenty rats were used in each experiment. Solutions were given intraperitoneally 15 min before exposure. The survival rate of control rats was 0–3 per cent.

As seen from the table, the prophylactic effect depends to a considerable extent on the composition of the salts. Thus, the lithium salt gave a percentage survival of rats more than twice as high as the sodium salt, although the total amount of SH- and NH$_2$-groups introduced into the body in the first case was 40 per cent less than in the second. An equivalent amount of Li, given in the form of acetate, did not give any protection. The remaining salts occupied an intermediate position as regards prophylactic effect. The observed differences in prophylactic effect of SH-glutathione salts are probably connected with the unequal access of protective material to the cell structures of the different tissues damaged by exposure. Additional confirmation for this was obtained in experiments on the rate of

TABLE 4

STRUCTURE AND PROPERTIES OF CERTAIN HYDROHALIDE SALTS OF ESTERS OF *l*-CYSTEINE WITH ALIPHATIC ALCOHOLS OF THE SERIES C_1–C_5

Salts of esters of *l*-cysteine	Chemical structure	Recrystallization from	Melting point, °C	Analysis. S content, %	
				Found	Calculated
Methyl ester hydrochloride	HS—CH₂—CH—COOCH₃ NH₂·HCl	Methyl alcohol	141	18·62	18·68
Ethyl ester hydrochloride	HS—CH₂—CH—COOC₂H₅ NH₂·HCl	Ethyl alcohol + ether	126–126·5	17·20	17·27
n-Propyl ester hydrochloride	HS—CH₂—CH—COOC₃H₇ NH₂·HCl	Chloroform + ether	76·5–77	15·94	16·05
n-Propyl ester hydrobromide	HS—CH₂—CH—COOC₃H₇ NH₂·HBr	Chloroform + ether	69·5–70	13·00	13·13
Isopropyl ester hydrochloride	HS—CH₂—CH—COOCH—CH₃ NH₂·HCl CH₃	Water containing HCl	153–153·5 (decomposition)	15·89	16·05

	Structure	Solvent	M.P.		
n-Butyl ester hydrochloride	$HS—CH_2—CH—COOC_4H_9$ $\quad\quad\quad\; NH_2 \cdot HCl$	Chloroform + ether	89–89.5	14.88	15.00
Isobutyl ester hydrochloride	$HS—CH_2—CH—COOCH_2—CH—CH_3$ $\quad\quad\quad\; NH_2 \cdot HCl \quad\quad\quad CH_3$	Isobutyl alcohol	92–92.5	14.85	15.00
Isoamyl ester hydrochloride	$HS—CH_2—CH—COOCH_2—CH_2—CH—CH_3$ $\quad\quad\quad\; NH_2 \cdot HCl \quad\quad\quad\quad\quad CH_3$	Ethyl acetate	82	13.96	14.08
Isoamyl ester hydrochloride	$HS—CH_2—CH—COOCH_2—CH—C_2H_5$ $\quad\quad\quad\; NH_2 \cdot HCl \quad\quad\quad CH_3$	Ethyl acetate	91–92	14.15	14.08
Isoamyl (tertiary) ester hydrochloride	$HS—CH_2—CH—COOC—CH_2—CH_3$ $\quad\quad\quad\; NH_2 \cdot HCl \quad CH_3 \quad CH_3$	Glacial acetic acid	162–164 (marked decomposition)	(13.00)	14.08

accumulation of non-protein SH-groups in the liver tissue of white rats, to which different salts of SH-glutathione were given (see p. 46).

Interest has been shown in studying the change in prophylactic properties as a result of substitution of the hydrogen of the carboxyl group by various organic radicals. For this purpose the synthesis and biological testing of salts of a number of esters of l-cysteine with aliphatic alcohols were carried out: the alcohols were methyl, ethyl, n-propyl, isopropyl, n-butyl, isobutyl, isoamyl, optically active isoamyl and tertiary isoamyl (amylene hydrate). Of these compounds the only ones described in the literature are the hydrochlorides of the methyl[53], ethyl[9], and benzyl esters[26]. There are no data in the literature on the prophylactic effect and biochemical changes of the methyl and ethyl esters of l-cysteine, introduced into the body in large doses. The hydrochlorides of l-cysteine esters were synthesized by the ordinary method (according to E. Fischer) used for preparing esters of simple amino acids. For one mole of chemically pure l-cysteine hydrochloride 5–7 moles of each of the corresponding alcohols were taken and dry hydrogen chloride was passed into the mixture for 1–1·5 hr with moderate heating. In a number of cases the separation of the ester salts presented considerable difficulties, but by using their different solubilities in various solvents all the substances could be obtained in good yields (not below 80 per cent). To obtain other salts of esters they were first converted into the free bases and were treated in ether solution with the corresponding acids. Salts of the ester disulphides were obtained by oxidizing the SH-form with H_2O_2 in a two-phase system (ether–water) in strictly controlled conditions. In Table 4 the constants of the compounds obtained are given.

Since the carboxyl group in esters is blocked, these compounds in their chemical properties are typical aminothiols. In addition to active SH- and NH_2-groups they contain an ester grouping and an aliphatic chain, straight or branched. The sulphydryl groups in the esters are oxidized to the —S—S—form both by molecular oxygen and by peroxide compounds, and the rate of oxidation depends to a great extent on the pH of the medium and the presence of catalysts (mainly traces of heavy metals: Fe, Cu, Mn). These esters are oxidized by alloxan, they add to quinones and react with alkylating agents, for example monoiodoacetic acid, nitrogen mustard, etc. They enter into reaction with carbonyl compounds (aldehydes, ketones, keto-acids, etc.), giving addition products of the semimercaptal type or hetero-cyclic compounds — derivatives of thiazolidine. They enter into reaction with purine and pyrimidine bases at the C=O groups. They react with carbon disulphide, giving complex S-derivatives of dithio-carbamic acid. Being strong bases, with mineral acids the esters form fairly stable, well crystallized salts, most of which possess high solubility in water and organic solvents (Table 5).

When the C atom nearest to the ester grouping is primary, the salts give aqueous solutions the pH of which is close to neutral. An aqueous solution of the isopropyl ester hydrochloride has an acid reaction. In this case, to reach pH 7 it is necessary to neutralize about 30 per cent of the hydrogen chloride contained in the salt. The consumption of alkali for neutralization of the hydrochloride of an ester with a tertiary C atom is considerably greater. When the solutions are at a neutral pH, some esters (n-propyl, n-butyl, isobutyl) are very stable as regards hydrolysis. The rate of hydrolysis was not studied at other pH values. The hydrohalide salts are stable on storage in the dry state: signs of decomposition (smell of alcohols) appear after 7–8 months; the hydrobromide of the n-propyl ester is more stable than the hydro-chloride. However, the most stable was the ethyl ester hydrochloride. The free esters (n-propyl, n-butyl, isobutyl) form stable salts also with a number of organic acids (some data on them will be given below.)

Esters of cysteine were tested as prophylactics on animals. In the preliminary experiments, the tolerated doses of these compounds were found. The results are given in Tables 6 and 7.

As seen from Table 6, a dose of 100 mg/100 g body weight is tolerated for almost all ester salts. An exception is the isoamyl ester, which is considerably more toxic than the others and even a dose of 50 mg/100 g gives an average increase in weight of less than 1 g/day. When introduced into the stomach, the doses of esters tolerated are considerably higher (up to 300 mg/100 g). In rats the visible symptoms after the introduction of sublethal doses of esters were distinct sluggishness, weakness, and ruffling of the fur. However, on the next day these symptoms completely disappeared, and later, in outward appearance, the animals in no way differed from normal ones. The administration of esters of lower alcohols (methyl, ethyl and propyl) to the rats in the tolerated doses had practically no effect on the rate of growth of the animals.

TABLE 5

RESULTS OF A STUDY OF THE SOLUBILITY OF CERTAIN
ORGANIC

l-Cysteine ester salts	Solvents				
	Water at 20°C, % by weight	Methyl alcohol	Ethyl alcohol	Ether	Acetone
Methyl ester–HCl	63·5	Soluble	Soluble	Insoluble	Sparingly soluble
Ethyl ester–HCl	52·7	Readily soluble	Soluble	Insoluble	Sparingly soluble
n-Propyl ester–HCl	86·1	Readily soluble	Readily soluble	Insoluble	Readily soluble
n-Propyl ester–HBr	90	Readily soluble	Readily soluble	Insoluble	Readily soluble
Isopropyl ester–HCl	45·2	—	—	—	—
n-Butyl ester–HCl	77·3	Readily soluble	Readily soluble	Insoluble	Sparingly soluble
Isobutyl ester–HCl	78·4	Readily soluble	Readily soluble	Insoluble	Sparingly soluble
Isoamyl ester–HCl		Readily soluble	Readily soluble	Insoluble	Soluble
Opt. active isoamyl ester–HCl	73·4	Readily soluble	Readily soluble	Insoluble	Readily soluble
Tertiary isoamyl ester–HCl	31·5	—	—	—	—

TABLE 5

l-CYSTEINE ESTER HYDROHALIDES IN WATER AND IN SOLVENTS

			Solvents			
Ethyl acetate	Acetic acid	Chloroform	Benzene	Petroleum ether	Carbon disulphide
Soluble	Soluble	Soluble	Insoluble	Insoluble	Insoluble
Soluble	Soluble	Sparingly soluble	Insoluble	Insoluble	Insoluble
Soluble	Readily soluble	Very readily soluble	Soluble	Insoluble	Insoluble
Soluble	Readily soluble	Very readily soluble	Soluble	—	—
—	—	—	—	—	—
Moderately soluble	Readily soluble	Readily soluble	Sparingly soluble	Insoluble	Insoluble
Soluble	Readily soluble	Readily soluble	Soluble	Insoluble	Insoluble
Readily soluble	Readily soluble	Readily soluble	Readily soluble	Insoluble	Insoluble
Soluble	Readily Soluble	Soluble	Soluble	Insoluble	Insoluble
—	—	—	—	—	—

TABLE 6

RESULTS OF A STUDY OF THE ACTION OF SALTS OF *l*-CYSTEINE ESTERS
ON NON-EXPOSED RATS WHEN SOLUTIONS WERE GIVEN
INTRAPERITONEALLY

l-Cysteine ester salts	Weight of rat, g	Dose of substance, mg/100 g	% of rats dead (toxicity)	Period of observation of surviving rats, days	Average weight of surviving animals, g	Average increase in weight, g/day
Control, uninjected	190	—	0	15	215	1·7
l-Cysteine, free	200	100	0	16	230	1·9
l-Cysteine, free	200	125	50	Not observed	—	—
l-Cysteine, free	190	150	100	—	—	—
Methyl ester–HCl	190	100	0	17	220	1·8
Methyl ester–HCl	190	150	67	17	185	0
Ethyl ester–HCl	180	100	0	15	210	2·0
Ethyl ester–HCl	175	150	0	11	197	2·0
Ethyl ester–HCl	170	200	100	—	—	—
n-Propyl ester–HCl	170	100	0	22	195	1·1
n-Propyl ester–HCl	180	125	50	22	200	0·9
n-Propyl ester–HBr	175	100	0	18	200	1·4
n-Propyl ester–HBr	200	123	0	18	220	1·1
n-Propyl ester–HBr	180	150	50	30	210	1·0
Isopropyl ester–HCl	185	100	0	16	205	1·3
n-Butyl ester–HCl	160	100	0	30	190	1·0
n-Butyl ester–HCl	200	150	83	10	210	1·0
Isobutyl ester–HCl	190	100	0	25	235	1·8
Isobutyl ester–HCl	175	150	30	30	215	1·3
Isoamyl ester–HCl	200	50	0	30	225	0·8
Isoamyl ester–HCl	220	100	100	—	—	—
Isoamyl (optically active) ester–HCl	170	50	0	30	200	1·0
Isoamyl (optically active) ester–HCl	190	80	0	30	220	1·0
Control, uninjected	190	—	0	15	215	1·7

TABLE 7

RESULTS OF A STUDY OF THE ACTION OF SALTS OF *l*-CYSTEINE ESTERS
ON NON-EXPOSED RATS WHEN SOLUTIONS WERE GIVEN ORALLY

l-Cysteine ester salts	Weight of rats, g	Dose of substance, mg/100 g	% of rats dead (toxicity)	Period of observation of surviving rats, days	Average weight of surviving animals, g	Average increase in weight, g/day
l-Cysteine, free	180	200	0	22	220	1·8
Methyl ester–HCl	190	200	0	17	240	2·9
Ethyl ester–HCl	175	200	0	11	195	1·8
n-Propyl ester–HCl	200	200	0	30	225	0·8
n-Propyl ester–HBr	200	200	0	15	220	1·3
Isopropyl ester–HCl	170	200	0	16	220	3·1
n-Butyl ester–HCl	190	200	0	15	225	2·3
Isobutyl ester–HCl	200	200	0	30	230	1·0
Isobutyl ester–HCl	185	300	0	16	185	0
Isoamyl (tertiary) ester–HCl	175	200	0	11	200	2·3

In dogs all the *l*-cysteine esters by any method of administration (particularly by the oral method) caused vomiting. Thus, in this respect the given esters did not differ from other prophylactics containing sulphur (cysteine, glutathione and salts of the simplest aminothiols and diaminodisulphides) and combinations of them with other substances. The only exception, first noticed in experiments on dogs, was a combination of the *n*-propyl ester hydrobromide with sodium nitrite (300 mg/kg; 18 mg/kg). After the intravenous administration of a mixture of these substances to dogs, no symptoms of vomiting were observed.

When the experiments were carried out on rats the solutions were given intraperitoneally 10–15 min before exposure (Table 8) and orally 30 min before exposure (Table 9). The dose in all cases was 650 r.

It is seen from Table 8 that many of the esters have a high prophylactic effect. The protective properties of the esters are characterized not only by a high percentage survival, but also by favourable characteristics in relation to changes in weight and to an increase in the average life span of animals dying later.

TABLE 8

RESULTS OF A STUDY OF THE PROTECTIVE ACTION OF SALTS OF
l-CYSTEINE ESTERS (INTRAPERITONEAL INJECTION)

Protective compound	Dose of substance, mg/100 g	SH-groups introduced, mg/100 g	Survival rate of rats at 30 days, numerator — those surviving, denominator — number of animals in experiment	Change in weight at 30 days, % of original	Average survival-time of rats that died, days
Control (without protective substance)	—	—	2/50	60	10
l-Cysteine, free (standard)	100	27·30	18/40	110	16
Methyl ester–HCl	100	19·27	14/20	116	19
Ethyl ester–HCl	100	17·81	17/32	105	12
n-Propyl ester–HCl	100	16·56	18/18	106	—
n-Propyl ester–HBr	123	16·65	18/18	108	—
Isopropyl ester–HCl	100	16·56	8/20	103	10
n-Butyl ester–HCl	100	15·47	9/15	97	16
Isobutyl ester–HCl	100	15·47	12/13	109	—
Isoamyl ester–HCl	50	7·26	12/20	—	4
Isoamyl (optically active) ester–HCl	80	11·62	13/18	105	14
Isoamyl (tertiary) ester–HCl	100	14·52	3/20	90	11

TABLE 9

RESULTS OF A STUDY OF THE PROTECTIVE ACTION OF CERTAIN
SALTS OF *l*-CYSTEINE ESTERS (ORAL ADMINISTRATION)

Protective compound	Dose of substance, mg/100 g	SH-groups introduced, mg/100 g	Survival rate of rats at 30 days, numerator — those surviving; denominator — number of animals in experiment	Change in weight at 30 days, % of original	Average survival-time of rats that died, days
l-Cysteine, free (standard)	200	54·60	10/20	105	14
Methyl ester–HCl	200	38·54	0/20	—	10
Ethyl ester–HCl	200	35·62	0/20	—	8·5
n-Propyl ester–HCl	200	33·12	2/16	110	9
n-Propyl ester–HBr	200	27·08	3/15	115	12
Isopropyl ester–HCl	200	33·12	6/20	89	12
Isobutyl ester–HCl	300	46·41	2/14	118	10

It is seen from the results in the tables that on transferring from one member of the series to another, no direct relationship is noted between the protective action and the amount of SH- and NH_2-groups introduced into the body. Evidently a number of factors play a part in the prophylactic effect, and primarily the structure of the molecule as a whole. There probably exists a more profound relationship between the prophylactic action, on the one hand, and the chemical, physico-chemical and radiation-biochemical properties of the prophylactics, on the other.

Like the salts of simple aminothiols, the salts of *l*-cysteine esters either give no protective effect at all or show a negligible effect, when introduced orally. It is suggested that these compounds scarcely enter the body from the gastro-intestinal tract. It should be noted that in experiments with oral administration of the solutions, the isopropyl ester gave the greatest prophylactic effect (30 per cent). This can be explained by the greater ease of hydrolysis of the ester group in this compound, which is accompanied by the formation of a certain amount of free *l*-cysteine. By the intraperitoneal route, of all compounds of this series the most effective were salts of the *n*-propyl ester of *l*-cysteine. On testing the hydrochloride of this ester at different exposure doses

TABLE 10

RESULTS OF A STUDY OF THE PROTECTIVE ACTION OF *l*-CYSTEINE
n-PROPYL ESTER WITH DIFFERENT EXPOSURE DOSES, AND ALSO IN
COMBINATION WITH OTHER SUBSTANCES

Protective compounds and combinations	Time between injection of solution and exposure, min.	Exposure dose, r	Survival rate of rats, %	Change in weight of surviving rats, % of original
n-Propyl ester–HCl (100 mg/100 g)	15	650	100	106
n-Propyl ester–HCl (100 mg/100 g)	15	700	100	100
n-Propyl ester–HCl (100 mg/100 g)	15	800	50	—
n-Propyl ester–HCl (100 mg/100 g) + ATP (7 mg/100 g)	15	650	100	103
n-Propyl ester–HCl (100 mg/100 g) + ATP (7 mg/100 g)	15	700	33	100
n-Propyl ester–HCl (100 mg/100 g) + NaNO$_2$ (5 mg/100 g)	40	800	35	100
n-Propyl ester–HCl (100 mg/100 g) + *d*-glucosamine (100 mg/100 g)	15	650	100	111
n-Propyl ester–HCl (100 mg/100 g) + *d*-glucosamine (100 mg/100 g)	15	700	100	103
n-Propyl ester–HCl (100 mg/100 g) + *d*-glucosamine (100 mg/100 g)	15	800	100	100
n-Propyl ester–HCl (100 mg/100 g) + *d*-glucosamine (100 mg/100 g)	15	900	17	76
n-Propyl ester–HCl (100 mg/100 g) + *d*-glucosamine (100 mg/100 g)	30	650	100	104
n-Propyl ester–HCl (100 mg/100 g) + *d*-glucosamine (100 mg/100 g)	60	650	100	106

(Table 10), the presence of a prophylactic effect was shown even at 800 r.

It is recorded that adenosine triphosphoric acid (ATP), which according to our results increases the protective properties of *l*-cysteine and glutathione, reduced the prophylactic action of the *n*-propyl ester at a dose of 700 r. On the other hand, *d*-glucosamine definitely increased the effectiveness of this ester (Table 10). In control experiments in the given conditions of exposure, glucosamine by itself did not protect rats and in combination with *l*-cysteine the effects of the latter did not increase.

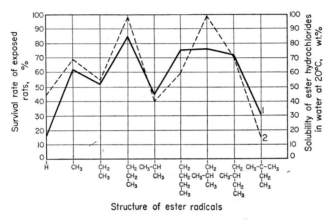

Structure of ester radicals

1 — solubility; 2 — survival rate

FIG. 1. Relationships between the structure and solubility of *l*-cysteine ester hydrochlorides and the survival rate of rats. The first point on the vertical axis corresponds to *l*-cysteine; the rest to esters of cysteine

In supplementary experiments, salts of the isobutyl ester of *l*-cysteine with thioglycollic and thiolactic acids were obtained in chemically pure form. Both salts dissolved easily in water and were studied as prophylactics on rats. The distinguishing property of these salts, particularly of the isobutyl ester thiolactate, was a strong soporific action. The animals fell into a deep and prolonged sleep even from small doses of the compound. However, when they were exposed in a condition of sleep after giving the most varied doses of these salts rich in free SH-groups, no prophylactic effect higher than that of the isobutyl ester hydrochloride was observed.

TABLE 11

RESULTS OF A STUDY OF THE RELATIONSHIP BETWEEN THE ACCUMULATION OF
NON-PROTEIN SH-GROUPS IN LIVER AND SPLEEN TISSUE OF RATS AND THE
SOLUBILITY OF *l*-CYSTEINE ESTER HYDROCHLORIDES

Protective compound	Solubility		Amount of substance injected, mg/100 g	SH-groups found, mg % (average for 6 rats)	
	in water at 20°C, wt. %	in chloro-form		in liver	in spleen
l-Cysteine, free	16·9	Insoluble	61	29·1 ± 0·6	15·6 ± 0·4
Methyl ester–HCl	63·5	Slightly soluble	86	35·7 ± 0·8	42·9 ± 1·0
Ethyl ester–HCl	52·7	Sparingly soluble	93	32·1 ± 0·8	34·7 ± 0·9
n-Propyl ester–HCl	86·1	Very readily soluble	100	44·8 ± 1·0	52·0 ± 1·1
n-Butyl ester–HCl	77·3	Readily soluble	107	30·6 ± 0·6	31·5 ± 0·8

Note—The experimental animals received the same dose of SH-groups (16·6 mg/100 g) in all cases.

In a number of *l*-cysteine ester hydrochlorides a distinct relationship was apparent between the protective properties of the compounds and their solubility in water. This relationship is shown in Fig. 1. It should be assumed that the prophylactics of this series, which have a higher solubility, also have increased ability to penetrate into the radio-sensitive tissues of the body. To confirm this theory a direct determination of the penetration of these compounds into the tissues was made. The degree of penetration was judged from the accumulation of non-protein SH-groups in the liver and spleen of white rats after the administration of certain ester hydrochlorides. The experiment was carried out in two variations. In one case, equimolecular doses of *l*-cysteine esters were introduced intraperitoneally. A protective dose of *n*-propyl ester hydrochloride, equal to 100 mg/100 g weight, was taken as standard; pH 6·9–7·1. The accumulation of SH-groups was determined in the liver and spleen 30 min after injection.

From the results in Table 11 a distinct relationship is seen between the solubility of the SH-compounds in water and organic solvents (chloroform) and the capacity of these thiols of raising the level of SH-groups in the tissues of the rat. After the administration of the esters, the concentration of SH-groups in the spleen tissue was higher than in the liver, whereas after the injection of free *l*-cysteine (61 mg/100 g) the reverse effect was observed.

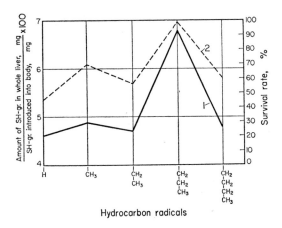

1 — amount of SH-groups; 2 — survival rate

FIG. 2. Relationship between the amount of non-protein SH-groups in the liver after injection of ester salts and the survival rate of rats

In the other case, "protective" doses of the compounds shown in Table 11 were given, and the amount of SH-groups in the same organs was also determined. The prophylactic action of these substances was determined in a parallel series of experiments.

The relationship between the accumulation of SH-groups in the liver and the survival rate of the rats is shown in Fig. 2. It is seen from this figure that the prophylactic effect changes in practically the same direction as the concentration of excess non-protein SH-groups in the liver tissue. *l*-Cysteine *n*-propyl ester hydrobromide is more soluble than the hydrochloride, and its prophylactic activity is higher.

Interest has been shown in studying the dynamics of the accumulation of non-protein SH-groups in liver and spleen tissues after

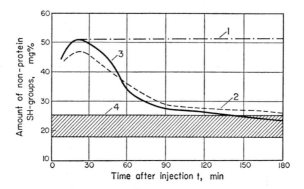

1 — *l*-cysteine, free (100 mg/100 g); 2 — *n*-propyl ester hydrochloride (100 mg/ 100 g); 3 — *n*-propyl ester hydrobromide (100 mg/100 g); 4 — biological variation in the normal

FIG. 3. Change in amount of non-protein SH-groups in liver of rats at different times after injection of *l*-cysteine and its *n*-propyl ester (hydrochloride and hydrobromide)

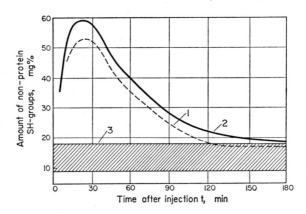

1 — *n*-propyl ester hydrochloride (100 mg/100 g); 2 — *n*-propyl ester hydrobromide (100 mg/100 g); 3 — biological variation in the normal

FIG. 4. Change in amount of non-protein SH-groups in spleen of rats at different times after injection of salts of *l*-cysteine *n*-propyl ester (hydrochloride and hydrobromide)

introduction of these salts into rats and in comparing the results obtained with the corresponding results for *l*-cysteine. The content of SH-groups was determined over a 3-hr period at different times after administration. It is seen in Figs. 3 and 4 that after the shortest intervals the level of SH-groups in the tissues (especially the spleen) rises sharply. The maximum levels of SH-groups in the tissues after injection of *l*-cysteine *n*-propyl ester hydrobromide are somewhat higher than after the introduction of the hydrochloride salt. Then a fairly rapid drop takes place and after 3 hr the amount of SH-groups in the organs approximates to normal. From Fig. 3 it is also seen that the character of the change of SH-groups in the liver tissue for esters differs strongly from that for *l*-cysteine. After injection of this amino acid the high level of SH-groups was maintained for the whole of the 3-hr period. The esters are excreted largely in the urine in the form of substances with non-oxidized SH-groups. A similar phenomenon was also noted for the simplest protective aminothiols, whereas after the introduction of cysteine (100 mg/100 g) the nitroprusside reaction in the urine was as a rule absent. It should be taken into account that with exposed rats (dose 650 r) to which the *n*-propyl ester hydrochloride was previously administered, the course of the curves of change of SH-groups (not shown on the figures) was practically the same as in non-exposed animals that had received the hydrochloride of the same ester. A similar effect was observed for aminothiols. As regards *l*-cysteine, the curves of the change of SH-groups are clearly different for non-exposed and exposed animals (see p. 59). These results indicate substantial differences in the biochemical (and radiation–biochemical) transformations of *l*-cysteine on the one hand, and its esters and simple aminothiols on the other.

One of the mechanisms which reduces the content of free *l*-cysteine of tissues is the reversible reaction of the biosynthesis of glutathione. Its SH-groups are probably also found in the non-protein filtrates long after the introduction of large doses of *l*-cysteine. For esters of *l*-cysteine (and simple aminothiols) this rapid method of detoxication probably does not occur and besides other metabolic reactions the elimination of non-oxidized sulphur from the body in the form of compounds with free SH-groups plays a considerable role. Probably certain other features of their biological action are connected with the peculiarities in the dynamics of the SH-groups of aminothiols and esters (rapid accumulation in the tissues and as much again the rapid disappearance of excess thiol groups). For possible mechanisms of the radioprophylactic action of aminothiol compounds see pages 50–53.

Besides the compounds indicated above, derivatives of thiazolidine-4-carboxylic acid with different substituents (R) in position 2 were synthesized and tested:

$$\begin{array}{c} \underset{2}{S}-\underset{1}{CH}-R \\ | \qquad | \\ \underset{5}{H_2C} \quad \underset{3}{NH} \\ \diagdown_4\diagup \\ CH-COOH \end{array}$$

The sulphur in these compounds is linked by a labile thioether link, which is easily split by hydrolysis with the formation of a free SH-group. The capacity of the C—S link for splitting depends to a large extent on the character of the substituents. All derivatives of thiazolidine carboxylic acid react with heavy metals (for example, mercury), forming mercaptides. The heterocyclic ring contains imino nitrogen, and the presence of the COOH group enables easily soluble salts to be obtained, which are suitable for injection into animals. Compounds of this series were synthesized mainly by Schubert's methods[47, 48]. In Table 12 different substituents (R) in position 2 are shown (from Nos. 2 to 17): 2-hydroxypropyl-, trichloromethyl-, phenyl-, 1-hydroxyphenyl-, 2-hydroxyphenyl-, 1-nitrophenyl-, 2-nitrophenyl-, 3-isopropylphenyl-, 2-methoxy-5-nitrophenyl-, 2,5-dimethoxyphenyl-, furyl-, and also residues of arabinose, xylose, galactose, glucose and glucosamine.

Most of the compounds of this series were ineffective in experiments on rats. The derivatives with phenyl and 2-hydroxyphenyl substituents gave a small prophylactic effect (about 30 per cent). Two compounds of this series (Table 12, Nos. 16 and 17) clearly showed the value of NH$_2$-groups in the side chain for the protective properties of the compounds. These substances, having the same heterocyclic nucleus, are very close in stereochemical configuration and differ only in the fact that one of them contains hydroxyl, and the other an amino group at the first carbon atom of the side chain. The hydroxyl derivative was ineffective, whereas the amino derivative was very active and gave 100 per cent survival rate of the animals which were tested. In order to

TABLE 12

RESULTS OF A STUDY OF THE RELATIONSHIP BETWEEN THE STRUCTURE
OF RADICALS IN POSITION 2 IN DERIVATIVES OF THIAZOLIDINE-4-
CARBOXYLIC ACID AND THE PROTECTIVE PROPERTIES OF THESE
DERIVATIVES

No. of derivative	Radical (R)	Literature references	Dose of substance mg/100 g	Number of rats in experiment	Survival rate after exposure, observed at 30 days
1	—H	(22, 44, 47)	20—50	12	Substance toxic
2	—CH$_2$—CH(OH)—CH$_3$	Not described in literature	75	18	Died from radiation disease from 8th to 20th day
3	—CCl$_3$	(47)	50	18	Three rats survived
4	—C$_6$H$_5$	(47)	100	18	Six rats survived
5	—C$_6$H$_4$ (OH) (1)		100	18	Six rats survived
6	—C$_6$H$_4$ (OH) (2)		50	10	All rats died from 7th to 14th day
7	—C$_6$H$_4$ (NO$_2$) (1)		50	12	All rats died from 8th to 12th day
8	—C$_6$H$_4$ (NO$_2$) (2)		100	12	All rats died from 9th to 21st day
9	—C$_6$H$_4$—CH$\diagup^{CH_3}_{\diagdown CH_3}$ (3)		25	12	All rats died on 2nd day
10	—C$_6$H$_3$ (OCH$_3$) (NO$_2$) (2,5)		50	10	Died from 7th to 14th day
11	—C$_6$H$_3$ (OCH$_3$) (OCH$_3$) (2,5)		50	10	Died from 7th to 14th day
12	—C$_4$H$_3$O	(47)	100	16	All rats died from 2nd to 15th day
13	$-\overset{\displaystyle OH}{\underset{\displaystyle H}{C}}-\overset{\displaystyle H}{\underset{\displaystyle OH}{C}}-\overset{\displaystyle H}{\underset{\displaystyle OH}{C}}-CH_2OH$	(48)	200	12	All rats died from 5th to 13th day

(continued)

TABLE 12 — *continued*

No. of derivative	Radical (R)	Literature references	Dose of substance mg/100 g	Number of rats in experiment	Survival rate after exposure, observed at 30 days
14	$\begin{array}{ccc} OH & H & OH \\ | & | & | \\ -C & -C-C- & -CH_2OH \\ | & | & | \\ H & OH & H \end{array}$	(48)	200	18	All rats died from 7th to 16th day
15	$\begin{array}{cccc} H & OH & OH & H \\ | & | & | & | \\ -C-C-C- & & -C-CH_2OH \\ | & | & | & | \\ OH & H & H & OH \end{array}$	(48)	200	12	All rats died from 6th to 23rd day
16	$\begin{array}{cccc} H & OH & H & H \\ | & | & | & | \\ -C-C-C- & & -C-CH_2OH \\ | & | & | & | \\ OH & H & OH & OH \end{array}$	(48)	200	12	All rats died from 7th to 15th day
17	$\begin{array}{cccc} H & OH & H & H \\ | & | & | & | \\ -C-C-C- & & -C-CH_2OH \\ | & | & | & | \\ NH_2 & H & OH & OH \end{array}$	Not described in literature	200	18	Survived more than 30 days. Weight on 30th day 104% of original

Note — The substances were given intraperitoneally 10–15 min before exposure (dose 650 r). Compounds Nos. 1–12 were dissolved in water with the addition of $NaHCO_3$, and substances Nos. 13–17 were dissolved in water without the addition of soda. In the controls the survival rate was 0–3 per cent and the duration of life 10–12 days in those that died.

interpret this relationship we carried out syntheses and tests on other substances of similar structure and, in particular, condensation products of prophylactic aminothiols and esters of *l*-cysteine with *d*-glucosamine.

In a subsequent series of experiments an attempt was made to find a link between the prophylactic action of sulphur-containing (and some other) compounds and their tendency to form stable complexes with metals. For this purpose compounds of the semi-mercaptal type were synthesized, namely: cysteinyl lactic and cysteinyl-β-hydroxypropionic acids, and also dihydroxymaleic acid (Table 13). These compounds were tested by L. I. Tikhonova (see p. 123) on an adsorption column and were shown to be effective complexing agents as regards yttrium and zirconium, approximating in stability of the resulting complexes to ethylenediaminetetraacetic acid, taken as standard.

In spite of the fact that certain of the compounds have low toxicity for animals (for example, the doses of cysteinyl lactic acid tolerated amounted to 2·5 g/1 kg weight), even with the greatest variations of doses and methods of administration, not one of them had a prophylactic effect, in contrast to *l*-cysteine esters, which have a low capacity for forming complexes.

TABLE 13

RESULTS OF A STUDY OF THE PROTECTIVE ACTION OF COMPOUNDS HAVING COMPLEX-FORMING PROPERTIES (DOSE 650 R)

Compound (acid)	Structure of compound	Reference to method of preparation	Protective effect on rats
Cysteinyllactic	CH_3—C—S—CH_2—CH—COOH with COOH above second C, OH below C, NH_2 below CH	(21, 47)	Not effective
Cysteinyl-β-hydroxy-propionic	CH_2—CH—S—CH_2—CH—COOH with COOH OH above, NH_2 below	Not described in literature	Not effective
Dihydroxymaleic	HOOC—C=C—COOH with OH OH below	(8)	Not effective

To study the radioprophylactic effect of chemical compounds, substances containing sulphur were synthesized and tested on animals. The effectiveness of prophylactics was compared with some of their chemical, physico-chemical and biological properties.

In addition the following was established:

1. Hydroxy- and ketoderivatives of mercaptopropionic acid did not protect rats from the action of X-rays.

2. The prophylactic effect of SH-glutathione depended to a considerable extent on the composition of its salts. In the case of the lithium salt the percentage survival rate of the rats exceeded that for the sodium salt by more than twice.

3. *l*-Cysteine esters with aliphatic alcohols of the C_1–C_5 series had a high prophylactic effect. The effectiveness of these compounds depended on the structure of the carbon chains, the composition of the salts and their solubility in water and organic solvents.

4. The relationship between the solubility of cysteine ester salts and the ability of these compounds to penetrate into radiosensitive tissues and also between the degree of accumulation of SH-groups in these tissues and the degree of prophylactic action, was shown.

5. The dynamics of the SH-groups in the liver and spleen tissue of rats after the administration of protective SH-compounds were studied. The results obtained indicate a substantial difference in the biochemical (and radiation-biochemical) transformations of cysteine on the one hand, and its esters and the simplest radioprotective aminothiols, on the other.

6. Sodium nitrite and ATP did not increase the prophylactic properties of *l*-cysteine *n*-propyl ester; *d*-glucosamine increased the effectiveness of this ester.

7. The prophylactic properties of derivatives of thiazolidine carboxylic acid were studied. Most of them were ineffective in experiments on rats. The value of the NH_2 group towards the prophylactic activity of a substance was made clear on a sample of two structurally similar compounds of this series.

8. In testing compounds of the semimercaptal type no connection was observed between the capacity of these substances to form (*in vitro*) stable complexes with metals and the ability to protect animals from the action of penetrating radiations.

9. A study of the structure, properties and biological action of compounds both possessing prophylactic activity and devoid of this capacity showed that the effectiveness of thiol substances depends on a definite chemical structure, the presence of additional functional groups in the molecule and their mutual disposition.

REFERENCES

1. ALEXANDER, P., BACQ, Z. M., COUSENS, S. F., FOX, M., HERVÉ, A. and LAZAR, J., *Radiat. Res.* **2**, 392 (1955).
2. BACQ, Z. M., *Advances in Radiobiol. (Lond.)*, 160 (1957).
3. BACQ, Z. M. and ALEXANDER, P., *Fundamentals of Radiobiology*, Butterworths, London (1955).
4. BACQ, Z. M., DECHAMPS, G., FISCHER, P., HERVÉ, A., LE-BIHAN, H., LECOMTE, J., PIROTTE, M. and RAYET, P., *Science* **117**, 633 (1953).
5. BACQ, Z. M. and HERVÉ, A., *Schweiz. Med. Wochenschr.* **82**, 1018 (1952).
6. BACQ, Z. M. and HERVÉ, A., *Strahlentherapie* **95**, 215 (1954).
7. BEILSTEIN's *Handb. d. Org. Chem.* **3**, 289 (1921).
8. BEILSTEIN's *Handb. d. Org. Chem.* **3**, 540 (1921).
9. BEILSTEIN's *Handb. d. Org. Chem.*, Zweites Ergw. **3/4**, 928 (1942).
10. BIILMANN, E., *Ann. d. Chem.* **339**, 351 (1905); *ibid.* **348**, 120 (1906).
11. BOND, V. P. and CRONKITE, E. P., *Ann. Rev. Physiol.* **19**, 229 (1957).
12. BRUES, A. M. and PATT, H. M., *Physiol. Rev.* **33**, 85 (1953).
13. CHAPMAN, W. H., SIPE, C. R., ELTZHOLT, D. C., CRONKITE, E. P. and CHAMBERS, F. W., *Radiology*, **55**, 865 (1950).
14. COHEN, J. A., VOS, O. and VAN BEKKUM, D. W., *Advances in Radiobiol. (Lond.)* 134 (1957).
15. CRONKITE, E. P., BRECHER, G. and CHAPMAN, W. H., *Proc. Soc. Exptl. Biol. Med.*, **76**, 396 (1951).
16. CRONKITE, E. P., BRECHER, G. and CHAPMAN, W. H., *Military Surgeon*, **109**, 294 (1951).
17. DEVIK, F., *Brit. J. Radiol.*, **27**, 463 (1954).
18. DOHERTY, D. G. and BURNETT, W. T., *Proc. Soc. Exptl. Biol. Med.*, **89**, 312 (1955).
19. ERDMANN, K., *Naturwissenschaften*, **40**, 147 (1953).
20. FISCHER, M. A., COULTER, E. P. and COSTELLO, M. J., *Proc. Soc. Exptl. Biol. Med.*, **83**, 266 (1953).
21. FRIEDMANN, E. and GIRSAVICIUS, J., *Biochem. J.*, **30**, 1886 (1936).
22. GENEVOIS, L. and CAYROL, P., *Bull. Soc. Chim.*, **6**, 1223 (1939).
23. GROUCH, B. G. and OVERMAN, R. R., *Science*, **125**, 1092 (1957).
24. HAGEN, U., *Advances in Radiobiol. (Lond.)*, **187** (1957).
25. HAJUDKOVIC, S. J. and KARANOVIC, J. I., *Bull. Inst. Nucl. Sci. (Belgrade)*, **7**, 139 (1957).
26. HARINGTON, C. R. and MEAD, T. H., *Biochem. J.*, **30**, 1598 (1936).
27. HERVÉ, A. and BACQ, Z. M., *Radiol. et Electrol.*, **33**, 651 (1952).
28. HOFMANN, D., *Strahlentherapie*, **96**, 396 (1955).
29. HOFMANN, D., KEPP, R. K., OEHLERT, G. and VASTERLING, H. W., *Strahlentherapie*, **96**, 1 (1955).
30. KHYM, J. X., SHAPIRA, R. and DOHERTY, D., *J. Amer. Chem. Soc.*, **79**, 5663 (1957).
31. KOCH, R., *Forsch. f. Geb. Röntgenstrahlen u. Nukl. Medizin*, **85**, 767 (1956).
32. KOCH, R., *Advances in Radiobiol. (Lond.)* 170 (1957).
33. KOELSCH, C. F., *J. Amer. Chem. Soc.*, **52**, 1105 (1930).
34. LAMBERT, G., MAISIN, J. and MANDART, M., *Compt. Rend. Soc. Biol.*, **146**, 1434 (1952).
35. LANGENDORFF, H., KOCH, R. and HAGEN, U., *Strahlentherapie*, **95**, 238 (1954).
36. LORENZ, W., *Strahlentherapie*, **88**, 190 (1952).
37. LANGENDORFF, H., KOCH, R. and HAGEN, U., *Strahlentherapie*, **100**, 137 (1956).

38. MacDonnel, L. R., Silva, R. B. and Feeney, R. E., *Arch. Biochem. Biophys.*, **32**, 288 (1951).
39. Meister, A., Fraser, P. E. and Tice, S. V., *J. Biol. Chem.*, **206**, 561 (1954).
40. Parrod, J., *Bull. Soc. Chim.*, **14**, 109 (1947).
41. Patt, H. M., *Physiol. Rev.*, **33**, 35 (1953).
42. Patt, H. M., Tyree, E. B., Straube, R. L. and Smith, D. E., *Science*, **110**, 213 (1949).
43. Patt, H. M., Smith, D. E., Tyree, E. B. and Straube, R. L., *Proc. Soc. Exptl. Biol. Med.*, **73**, 18 (1950).
44. Rather, S. and Clarke, H. T., *J. Amer. Chem. Soc.*, **59**, 200 (1937).
45. Rugh, R. and Wang, S. C., *Proc. Soc. Exptl. Biol. Med.*, **83**, 411 (1953).
46. Schneider, F. and Reinefeld, E., *Biochem. Z.*, **318**, 507 (1948).
47. Schubert, M. P., *J. Biol. Chem.*, **114**, 341 (1936).
48. Schubert, M. P., *J. Biol. Chem.*, **130**, 601 (1939).
49. Shapira, R., Doherty, D. G. and Burnett, W. T., *Radiat. Res.*, **7**, 22 (1957).
50. Smith, D. E., Patt, H. M., Tyree, E. B. and Straube, R. L., *Proc. Soc. Exptl. Biol. Med.*, **73**, 198 (1950).
51. Storaasli, J. P., Rosenberg, S. and Weisberger, A., *Nucl. Sci. Abstr.*, **8**, 694 (1954).
52. Struabe, R. L. and Patt, H. M., *Proc. Soc. Exptl. Biol. Med.*, **84**, 702 (1953).
53. Sullivan, M. X., Hess, W. C. and Howard, H. W., *J. Wash. Acad. Sci.*, **32**, 285 (quoted from *Chem. Abstr.* 1943, 37, 900) (1942).

Mechanism of the Protective
Action of Certain Thiol Compounds

V. G. Yakovlev and L. S. Isupova

THERE is an extensive literature on the protective action of chemical compounds, and much empirical material has been accumulated, but the mechanism of chemical protection has been studied little.

In experiments on mice and rats, into which protective materials have been injected not long before exposure, many authors have succeeded in obtaining a fairly high percentage survival rate with an almost total loss of the control animals[1, 3, 15]. If it is considered that previously no measures had been taken to increase the resistance of the animals to radiation, and after exposure no elements of therapy were used for the radiation sickness, then there are sound reasons for thinking that protective substances, as such, actually decrease the severity of radiation sickness.

Among the compounds which possess protective action, certain thiol substances with a definite "protective" structure offer the greatest interest.

The present investigation is devoted to a study of the mechanism of action of certain nitrogen compounds containing sulphur, known for their capacity for protecting animals from the effects of penetrating radiation during external total exposure to X-rays in lethal doses.

Considerable interest was offered by the study of the relationship between the degree of protective action and the concentration of low molecular-weight SH-compounds in the tissues after the introduction of certain sulphur-containing compounds into the body of an animal. The substances used included both well known and new ones, containing free SH-groups, disulphide and thioether links, and also isothiouronium groupings. The concentration of low molecular-weight thiol substances in the tissues was judged from the amount of titrated non-protein SH-groups in a non-protein filtrate. A relationship was found between the survival rate of the tested animals and the concentration of excess SH-groups in the tissues both during the testing of certain thiolic protective compounds closely related in structure, and also for substances containing different types of non-oxidized organic sulphur linkages. In addition, for each group of compounds mechanisms were considered. The method of administration and rate of penetration of the protective substances into the organs was of great importance.

The experiments were carried out on male white rats (180–220 g), which were maintained on an ordinary diet. The results of two parallel series of experiments were compared: (a) on the determination of SH-groups in the organs (liver) in non-exposed animals and (b) on the protective action.

1. Rats were decapitated at different times after the intraperitoneal or oral introduction of solutions of the substances being studied (1 ml/100 g weight). The liver was quickly homogenized in ice with a 4 per cent solution of sulphosalicylic acid to precipitate the albumens with the addition of sodium ethylenediaminetetraacetate as stabilizer, which protects the SH-groups from oxidation in the presence of traces of heavy metals. In 5 ml. proteinless filtrate (total volume 40 ml.) the sulphydryl groups were determined by the method of back titration using p-chloromercuribenzoic acid. The average values for each variation were calculated from experiments on 4–10 rats. For each experiment a control was set up to determine the endogenous non-protein SH-groups of the tissues being studied in non-exposed animals, the content of which for the liver varied from 18 to 25 mg per cent depending on the group of animals, the season and other conditions. The excess SH-groups were determined by the difference between the total amount of SH-groups in the non-protein filtrates of the liver tissue of the experimental and control animals.

2. The control and test ("protected") rats were exposed under identical conditions to X-rays in a RUM-3 apparatus at doses of 650 r, close to the minimum 100 per cent lethal dose. At this dose the loss of control animals was 95–100 per cent. The solutions of protective substances were introduced in the same doses and concentrations as in series "a" 15–30 min before exposure.

Conditions of exposure: voltage 180 kV; current strength 15 mA; dose-rate 40 r/min; focal length 40 cm; filter 0·5 mm Cu + 1 mm Al; two rats were exposed in Plexiglass boxes with lattice walls. The criteria of protective action were the survival rate 30 days after exposure and the change in weight of the animals. The number of rats in the individual experiments was 8–10; the number of experiments

TABLE 1

RESULTS OF A STUDY OF THE PROTECTIVE ACTION ON RATS AND OF THE LEVEL OF NON-PROTEIN SH-GROUPS IN THE LIVER FOLLOWING INTRAPERITONEAL INJECTION OF *l*-CYSTEINE DERIVATIVES

Protective compound	Chemical structure	Dose of substance, mg/100 g	Sulphur introduced, mg/100 g	Sulphur of SH-groups found, % of normal	Excess of sulphur of SH-groups, % of that introduced	Survival rate of animals, %
l-Cysteine (free)	HS—CH$_2$—CH(NH$_2$)COOH	100·0	26·5	232·5	2·42	40—70
n-Propyl ester (HCl salt)	HS—CH$_2$—CH(NH$_2$)COO× ×C$_3$H$_7$·HCl	100·0	16·1	224·0	3·74	60—100
n-Propyl ester (HBr salt)	HS—CH$_2$—CH(NH$_2$)COO× ×C$_3$H$_7$·HBr	122·3	16·1	237·5	4·14	80—100
Disulphide of *n*-propyl ester (dihydro-bromide)	S—CH$_2$—CH(NH$_2$)COO× ×C$_3$H$_7$·2HBr | S—CH$_2$—CH(NH$_2$)COO× ×C$_3$H$_7$	120·0	15·8	136·7	1·18	10—20
S-β-Hydroxy-ethyl-*l*-cysteine (free)	S—CH$_2$CH(NH$_2$)·COOH | CH$_2$—CH$_2$OH	100·0	19·4	104·0	0·12	0

Note—In the last column but one of Table 1 the excess non-protein sulphur of SH-groups accumulated by the whole liver is shown.

C

for each substance was different, but not less than three. The average values are given in the tables. The variation of the figures did not exceed 5–8 per cent.

Data on the content of non-protein SH-groups in the liver of rats when protective substances containing sulphur were injected are given below. The results of determinations of non-protein SH-groups are given for a single time-interval — 30 min after the injection of the substances being investigated. For a large number of the experiments using one protective substance or another the content of SH-groups reached maximum values by this time. In the experiments with exposure the 30 min consisted of the interval between the injection and the exposure and the duration of the exposure itself.

In Table 1 results are given for *l*-cysteine and its derivatives: *l*-cysteine *n*-propyl ester hydrochloride and hydrobromide, the dihydrobromide of *l*-cysteine *n*-propyl ester disulphide and S-β-hydroxyethylcysteine. The substances were given intraperitoneally in "protective" doses, which were found previously for each compound by experiment. In a number of cases these doses approached the maximum that could be tolerated. Thus, as each compound was tested, different amounts of sulphur were introduced into the body.

It is seen from the results in Table 1 that the prophylactic effect is in direct relation to the concentration of SH-groups in the rat liver tissue.

In the case of salts of cysteine propyl ester, which are more effective than *l*-cysteine itself (see present symposium, p. 24), it is characteristic that the quantity of SH-groups introduced is less, but their accumulation in the liver tissue is more intense. This is apparently explained by the greater ability to penetrate of the *l*-cysteine ester. This is especially true of its hydrobromide (see p. 29) and agrees with the results obtained in special experiments with ^{35}S cysteine and ^{35}S cysteine *n*-propyl ester (hydrobromide). The radioactivity of the liver and spleen tissues was determined 30 min after intraperitoneal injection into rats of the indicated doses of the given substances (6000 counts/min.g weight) in a mixture with "protective" quantities of the same non-radioactive compounds. The results of the determinations indicate parallelism between the accumulation of SH-groups and the concentration of ^{35}S. Thus, in liver tissue, in the case of labelled cysteine 79 counts/min/10 mg tissue was found, and for the *n*-propyl ester–HBr — 121 counts/min/10 mg tissue. As regards the spleens, in this case ^{35}S cysteine penetrates into the tissue somewhat more intensely (32 counts/min/10 mg tissue) compared with the radio-active sulphur of cysteine n-propyl ester (22 counts/min/10 mg). l-Cysteine n-propyl ester disulphide gives a relatively small increase of SH-groups, and its protective action does not exceed 20 per cent. At the same time, low molecular-weight SH compounds appear, possibly both as a result of direct reduction according to the scheme: R—S—S—R $\rightleftarrows$ 2 SHR[13], and as a result of the interaction of the disulphide with the SH-groups of the proteins according to the scheme[6] protein —SH+R—S—S—R $\rightleftarrows$ protein —S—S—R+HS—R.

TABLE 2

RESULTS OF A STUDY OF THE PROTECTIVE ACTION ON RATS AND OF
THE LEVEL OF NON-PROTEIN SH-GROUPS IN THE LIVER WHEN
CYSTEAMINE DERIVATIVES ARE INJECTED INTRAPERITONEALLY

Protective compound	Chemical structure	Dose of substance, mg/100 g	Sulphur introduced, mg/100 g	Sulphur of SH-groups found, % of normal	Excess sulphur of SH-groups, % of that introduced	Survival rate of animals, %
Cysteamine (HCl salt)	HS·CH$_2$·CH$_2$·NH$_2$·HCl	15	4·23	121·0	2·29	~40
Cysteamine (HBr salt)	HS·CH$_2$·CH$_2$·NH$_2$·HBr	15	3·04	129·0	4·39	60
S-β-Amino-ethyliso-thio-uronium Br.HBr	HBr·NH$_2$·CH$_2$·CH$_2$·S—C $\diagdown$ NH / NH$_2$·HBr	25	2·85	118·0	2·98	25
S-β-Amino-ethyliso-thio-uronium Br.HBr	HBr·NH$_2$·CH$_2$·CH$_2$·S—C $\diagdown$ NH / NH$_2$·HBr	50	6·84	144·0	3·12	0

Finally, the last compound, in which the sulphur is linked by a stable thioether bond, does not liberate SH-groups in the tissues and has no prophylactic effect.

In Table 2, results are given for aminothiols and their derivatives. Hydrochlorides and hydrobromides of cysteamine (β-mercapto-ethylamine) and the hydrobromide of S-β-aminoethylisothiouronium bromide were studied.

It is seen from Table 2 that all the substances studied increase the level of SH-groups and have pronounced prophylactic properties. On comparing the effects of the salts of individual aminothiols it is seen that the hydrobromide salts are more effective and bring about a greater increase in concentration of SH-groups than the hydro-chlorides. This may also be the result of unequal penetration of these salts into the cellular structures.

It is very interesting that aminoethylisothiouronium, while possessing a prophylactic effect, also raises the level of the non-protein SH-groups. Evidently conditions may occur in the tissues for the hydrolytic breakdown of the isothiouronium compounds according to the following scheme[11]:

$$
\begin{array}{c}
R_1 \\ \diagdown \\ \qquad N-CH_2-CH_2-S-\overset{\displaystyle H|OH}{\underset{\big\downarrow}{C}}\diagup\!\!\diagup^{NH}_{\diagdown NH_2} \\ R_2 \diagup \\[2mm]
\big\downarrow \\[2mm]
R_1 \\ \diagdown \\ \qquad N-CH_2-CH_2-SH + O=C\diagup^{NH_2}_{\diagdown NH_2} \\ R_2 \diagup
\end{array}
$$

The increase in content of SH-groups in the tissues after injection of aminoethylisothiouronium salts may also be the result of an intra-molecular transguanylation reaction with the formation of β-mercapto-ethylguanidine according to the equation[4, 10, 14]:

$$
H_2N-CH_2-CH_2-S-C\diagup^{NH}_{\diagdown NH_2} \longrightarrow HS-CH_2-CH_2-NH-C\diagup^{NH}_{\diagdown NH_2}
$$

When the dose of aminoethylisothiouronium is increased to 50 mg/100 g the content of SH-groups is increased, but a toxic effect becomes apparent.

In Table 3, data are given concerning the prophylactic effect and content of non-protein SH-groups in the liver after the oral administration of cysteine and cysteamine derivatives to rats.

To compare the effect of these substances with the effect of free *l*-cysteine, the same amount of sulphur calculated at 53 mg/100 g body weight was in all cases introduced into the body of the animal as

TABLE 3

RESULTS OF A STUDY OF THE PROTECTIVE ACTION ON RATS AND OF THE LEVEL OF NON-PROTEIN SH-GROUPS IN THE LIVER WHEN DERIVATIVES OF *l*-CYSTEINE AND CYSTEAMINE ARE GIVEN ORALLY

(S, 52·9 mg/100 g introduced)

Protective compound	Dose of substance, mg/100 g	Sulphur of SH-groups found, % of normal	Excess sulphur of SH-groups, % of that introduced	Toxicity of dose, % mortality	Survival rate, %	
					With given dose of sub-stance	With smaller doses
l-Cysteine, free	200	173·3	0·70	0	about 70	—
n-Propyl ester (HCl salt)	329·6	107·9	0·08	0	0	0
n-Propyl ester (HBr salt)	403·0	149·4	0·56	30	—	~30
Cysteamine (HBr salt)	260	134·3	0·36	100	—	~20
S-β-Aminoethyl-isothiouronium Br.HBr	464	189·1	0·93	100	—	Not observed

part of the composition of these compounds. This dose was non-toxic only for *l*-cysteine and the hydrochloride of its *n*-propyl ester. As a result of the toxicity of such doses, the remaining substances gave from 30 to 100 per cent loss of animals. However, when used in smaller (tolerated) doses, for example, *l*-cysteine *n*-propyl ester HBr at 200–250 mg/100 g, they had a prophylactic effect of the order of

20–30 per cent, the degree of which also depended on the level of SH-groups in the tissues. There is a characteristically sharp difference in the effect of *l*-cysteine and of the hydrochloride of its *n*-propyl ester both in relation to the prophylactic effect and the influence on the level of SH-groups in the liver tissue. Evidently salts of aminothiols (and of *l*-cysteine esters) hardly ever enter the body in unchanged form from the gastro-intestinal tract (see p. 25).

TABLE 4

RESULTS OF A STUDY OF THE PROTECTIVE ACTION ON RATS AND OF THE LEVEL OF NON-PROTEIN SH-GROUPS IN THE LIVER WHEN SALTS OF SH-GLUTATHIONE AND DERIVATIVES OF *d,l*-METHIONINE ARE GIVEN

Protective compound	Method of administration	Dose of substance, mg/100 g	Sulphur introduced, mg/100 g	Sulphur of SH-groups found, % of normal	Excess sulphur of SH-groups, % of that introduced	Survival rate of rats, %
SH-Glutathione (Na salt)	Intraperitoneal	250	26·07	169·0	1·32	20—30
SH-Glutathione (Li salt)	Intraperitoneal	250	26·07	193·0	1·77	70 (dose of substance reduced)
d,l-Methionine	Intraperitoneal	65	13·97	108·2	0·35	0
n-Propyl ester HCl	Intraperitoneal	200	28·16	98·6	—	0
n-Propyl ester HCl	Oral	374	52·66	104·2	0·04	0

In supplementary experiments the effect of substances of the "cysteine-cysteamine group" on the concentration of non-protein SH-groups in the spleen and testes of white rats was studied, because the tissues of these organs are some of the most sensitive to the action of radiation. After the introduction of the SH-compounds into the body, the change in content of SH-groups in the spleen took place in approximately the same direction as in the liver tissue. The degree of penetration of the SH-groups of *l*-cysteine into the tissue of the testes was considerably less than into the liver tissue. For cysteamine this difference is less clearly pronounced.

In Table 4 the prophylactic action of glutathione salts and their effect on the level of SH-groups in liver is shown.

It is seen from the results given that the Li salt is considerably more effective than the Na salt in prophylactic action and in ability to penetrate. The latter may be judged from the large accumulation of SH-groups in the liver in the lithium salt test. Cations may to a large extent affect the penetration of the salts of prophylactics (see p. 15).

In the same table, results are given for *d,l*-methionine and its *n*-propyl ester, which in the form of the hydrochloride salt is extremely soluble in water and in organic solvents, and is fairly easily tolerated by animals even in large doses when given intraperitoneally and orally. However, neither methionine itself nor its ester possess protective properties. It is possible that the biological demethylation of methionine with the formation of homocysteine (which has a prophylactic action)[12] takes place slowly and to a negligible extent. It should

TABLE 5

RESULTS OF A STUDY OF THE LEVEL OF NON-PROTEIN SH-GROUPS IN THE LIVER OF RATS WHEN "INHIBITORS" AND "STIMULATORS" OF THE PROTECTIVE ACTION OF *l*-CYSTEINE ARE GIVEN INTRAPERITONEALLY

Compound	Dose of substance, mg/100 g	Amount of SH-groups, % of normal	Survival rate of rats, %
l-Cysteine	100	210	40–70
Pyruvic acid	89	100	0
Oxalacetic acid	130	100	0
ATP	30	100	0
l-Cysteine+pyruvic acid	100+89	165	15–20
l-Cysteine+oxalacetic acid	100+130	162	15–20
l-Cysteine+ATP	100+10	240	70–100

also be borne in mind that *in vitro* the thioether sulphur of methionine (and its ester) is easily oxidized by peroxidic compounds to the corresponding sulphoxide[17]. The possibility of this process also occurring *in vivo* is not excluded.

In Table 5 the change in content of SH-groups after intraperitoneal injection of *l*-cysteine into the rat together with "inhibitors" and "stimulators" of its prophylactic action is shown.

As seen from Table 5, the biogenic "inhibitors" pyruvic and oxalacetic acids decrease the amount of excess SH-groups in the liver almost two-fold and sharply reduce the prophylactic effect of *l*-cysteine. Adenosine triphosphoric acid (ATP) in combination with *l*-cysteine increases the level of excess SH-groups in the liver and increases the prophylactic effect. It is possible that in this case a partial phosphorylation of the *l*-cysteine takes place with the formation of S-phosphocysteine, which is more reactive and easily penetrates through tissue membranes. At the same time adenosine monophosphoric acid does not exert any influence on the protective properties of *l*-cysteine.

TABLE 6

RESULTS OF A STUDY OF THE EFFECTS OF MERCAPTOACETIC ACID AND "UNITHIOL" ON THE LEVEL OF SH-GROUPS IN THE LIVER AND ON THE SURVIVAL RATE OF RATS

Compound	Chemical structure	Dose of substance, mg/100 g	SH-groups introduced, mg/100 g	SH-groups found, % of normal	Protective action
"Unithiol" (Na salt)	CH_2—CH—CH_2—SO_3Na $\|$ $\|$ SH SH	50	15·7	100	Ineffective
"Unithiol" (Na salt)	CH_2—CH—CH_2—SO_3Na $\|$ $\|$ SH SH	100	31·4	162·5	Toxic
Mercaptoacetic acid (Na salt)	HS—CH_2—COONa	94	27	225	Ineffective

Investigations are being carried out on the phosphoric esters of SH-aminoacids and aminothiols.

In Table 6 the results on the testing of "unithiol" and mercapto-acetic acid for prophylactic action are given.

As seen from the results given, small doses of "unithiol" (50 mg/100 g weight) gave neither a prophylactic effect nor an accumulation of excess SH-groups in the liver. With larger doses (100 mg/100 g weight) the excess SH-groups appear, but a toxic effect of the substance is noted. Further investigations are necessary to confirm the conclusions reached, because the results of different authors on this problem are contradictory.

Mercaptoacetic acid is as it were an exception to the rule which we observed: according to the results of a number of authors its redox potential is equal to that for *l*-cysteine[8, 9]. When it is used, a considerable accumulation of excess SH-groups is observed in the liver, and at the same time it is not a protective substance in experiments *in vivo*. In this case the necessity for the presence of NH_2 groups and the significance of the general structure for prophylactic action is indicated particularly clearly.

It should be noted that a relationship between the prophylactic action and the accumulation of SH-groups in tissues occurs only in very limited groups of protective compounds. In the experiments carried out there were cases where it was not possible to show any effect of sulphur-containing compounds on the level of SH-groups in the liver, although these compounds had a clearly pronounced prophylactic action. Of these compounds, it is possible to note for example, 2-amidinothiourea

$$\left(\begin{array}{c} \overset{NH}{\underset{\parallel}{}} \quad \overset{S}{\underset{\parallel}{}} \\ NH_2-C-NH-C-NH_2 \cdot HCl \end{array} \right)$$

It does not give an accumulation of excess SH-groups, but has prophylactic properties (see p. 78). At the same time there are direct indications of the significance of the SH-groups in the protective effect. Compounds in which the SH-group is replaced by an OH-group are quite ineffective, for example, serine (the hydroxyl analogue of cysteine) and monoethanolamine, the analogue of β-mercaptoethyl-amine.

Protective thiol compounds, introduced into the body as such, or formed in it from other substances, are effective only in the presence of a certain molecular structure. For example, lengthening the carbon chain in aminothiols to more than three C atoms reduces the protective action of these compounds[11]. Rearrangement of the functional groups in some cases is also of substantial significance. As regards the mechanism of the protective action of thiol compounds, not one of the proposed theories at present offers an exhaustive explanation. Some investigators suggest that the protective thiol compounds, being antioxidants, reduce the oxygen tension in the tissues. This decreases the formation of oxidizing radicals, the presence of which causes subsequent chain reactions in the tissues, leading to the development of radiation disease. However, there is as yet no experimental proof that after the introduction of the protective substances into the body the content of dissolved molecular oxygen in the tissue liquids is decreased. However, there is, seemingly, some indirect evidence: some protection by thiols against the toxic effect of oxygen under pressure and the positive effect of anoxia at the moment of exposure may be accounted for by quantitative changes in the protective substances in the body. Thus, in rats with increasing oxygen starvation, the content of non-protein thiol compounds, which may play the role of protective substances, increases sharply in the tissues (for example, in the spleen tissue almost two-fold).

Protective thiol compounds introduced into the body are as a rule effective in the maximum tolerated doses, i.e. when the exposure takes place in the presence of pronounced toxic symptoms: in rats, sluggishness, a reduction in the consumption of oxygen, and the appearance of a large amount of SH-groups in the urine are observed; in dogs, vomiting, diarrhoea, disturbance of the respiration and cardiac activity. However, these symptoms pass quickly and without trace. When the body is loaded with protective thiols, biological mechanisms for rendering these substances harmless come into operation. These mechanisms are oxidation and excretion. However, when there is a large excess of an SH-compound introduced, a third method arises which may be called "buffering" of the thiol, i.e. as a result of reversible reactions the formation of temporary addition products of the protective substances with biomolecules. It is possible that it is just these reactions which play an essential part in the protection of the biomolecules from the action of radiations. They may take place in the following principal ways:

1. By forming short-lived mixed disulphides with the SH-groups of the tissues. Eldjarn and co-workers[5-7] hold to this point of view, suggesting the temporary screening of the radiation-sensitive SH-groups of the protein structures. We obtained interesting results during the study of the effect of protective substances on the protein SH-groups of rat liver. Thus, 30 min after the intraperitoneal injection of l-cysteine into rats in protective doses (100 mg/100 g) the content of the total protein SH-groups in the liver was reduced by about 17 per cent in comparison with their level in healthy animals.

It is possible that this reduction in the natural protein SH-groups takes place because of the formation of mixed disulphides of the protein with the l-cysteine introduced into the body in an excess quantity.

2. By adding thiols to double bonds. This mechanism may play a part in preventing secondary reactions in irradiated unsaturated aliphatic acids. This question has not yet been studied.

3. By forming complexes with the metals of enzyme systems and inhibiting the action of a number of oxidases[2]. This has not been studied in experiments *in vivo*.

4. Protective thiol compounds may enter into reversible reactions with certain $>C=O$ groups, which enter into the composition of many biologically important compounds. The new atomic groupings or heterocyclic rings thereby formed may themselves during exposure take up part of the absorbed energy, and the atomic groupings and bonds of the biomolecules will be to some extent protected from the direct or indirect action of radiation. The possibility is not excluded that the temporary addition of the aminothiols to aldehyde groups existing in the tissues protects these groups from oxidation, caused by irradiation, to carboxyl groups.

We noted that many compounds containing reactive $>C=O$ groups (ketoacids, sugars, aliphatic aldehydes and so on), when introduced into the body, not only do not possess protective properties but even inhibit the action of protective thiolic substances (see Table 5), evidently by entering into reaction with them. In most cases the synthetically obtained condensation products of aminothiols with aldehydes and ketones do not have a prophylatic effect (see p. 32).

It is possible that the protection mechanism under consideration (addition of protective substances to various $\diagdown C{=}O$ groups) is of substantial significance. Among biologically important compounds having reactive $\diagdown C{=}O$ groups, it is sufficient to indicate certain co-enzymes, for example, pyridoxal and O,S-diacetylthiamine[16],

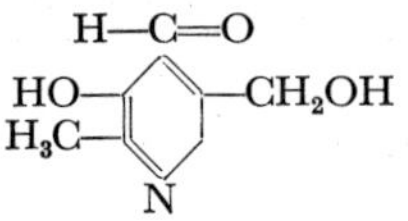

pyridoxal

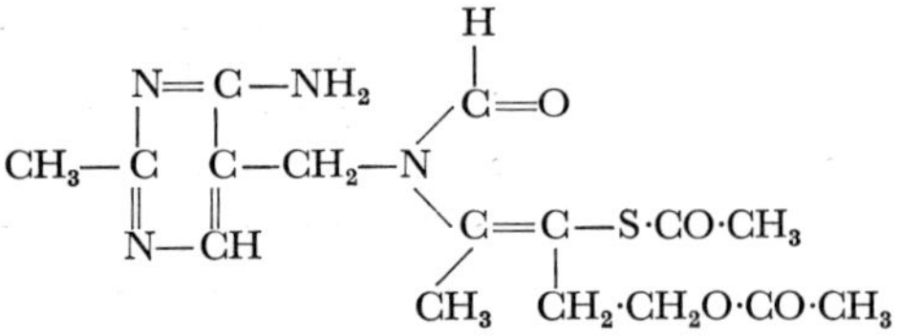

O,S-diacetylthiamine

and also the structural elements of nucleic acids, for example, guanine and uracil.

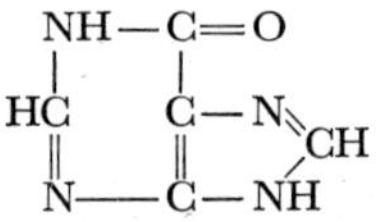

guanine
(ketoform)

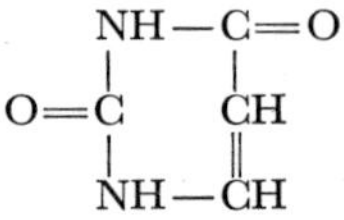

uracil
(ketoform)

Aminothiolic substances can enter into reaction with $\diagdown C{=}O$ groups in two ways: (1) with the formation of semithioacetals and (2) with the formation of new heterocyclic nuclei, for example, thiazolidine (with local changes of pH).

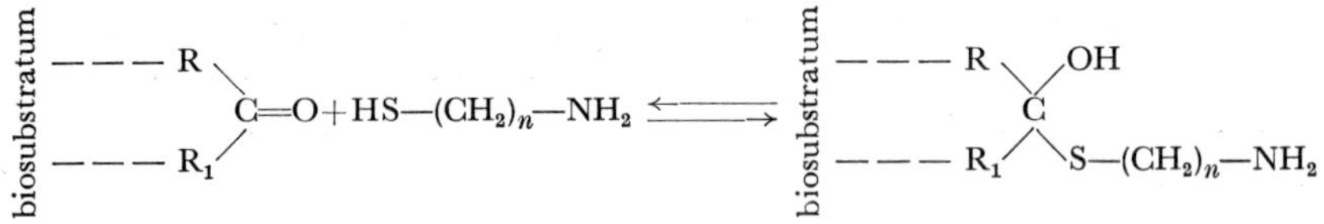

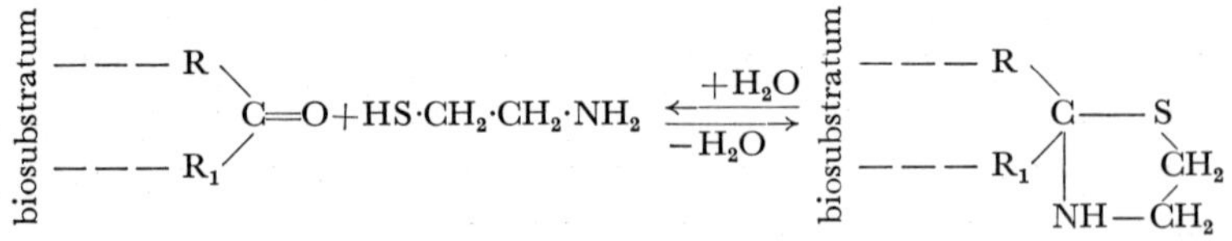

In experiments *in vitro* crystalline addition products of certain aminothiols with guanine and uracil have been obtained. Their composition will be studied. The possibility is not excluded that the prophylactics protect the molecules of purine and pyrimidine bases, by the method indicated above, from rearrangements caused in the heterocyclic nuclei by irradiation.

It should be considered that in blocking a $>C=O$ group, for example in a uracil residue, this part of the molecule may temporarily lose the capacity for enolization, i.e. for the formation, in the ring, of the chromophore $-C=C-C=N-$, which is responsible for the absorption of secondary electromagnetic radiation. In consequence, changes in nucleic acids, occurring under the influence of radiation, may be less pronounced during chemical protection.

In addition it must be noted that, assuming the formation *in vivo* of heterocyclic rings from aminothiols, it can be explained why lengthening of the carbon chain and rearrangement of the functional groups in chemical compounds has such a strong effect on their protective action. Increasing the number of C atoms in the chain to more than three would theoretically be accompanied by the formation of seven-membered rings. However, such rings are unstable and are probably not formed in the body.

The role of functional groups is well illustrated by the following example. The presence of an electronegative carboxyl group in the vicinity of the sulphur atom in isocysteine completely abolishes the protective action of this substance.

$$HS-CH-CH_2-NH_2-\text{cysteamine—effective}$$
$$|$$
$$H$$

$$HS-CH-CH_2-NH_2-\text{isocysteine—not effective}$$
$$|$$
$$COOH$$

The protection of the body from ionizing radiations at the present stage is acquiring still greater importance. Accordingly the theories put forward in the present work concerning the mechanism of the protective action of thiol compounds should attract the attention of biochemists and radiobiologists working in this field.

REFERENCES

1. ALEXANDER, P., BACQ, Z. M., COUSENS, F., FOX, M., HERVÉ, A. and LAZAR, J., *Radiat. Res.*, **2**, 392 (1955).
2. BERSIN, T., quoted from *Chem. Abstr.*, **32**, 5006 (1938).
3. DOHERTY, D. G. and BURNETT, W. T., *Proc. Soc. Exptl. Biol. Med.*, **89**, 312 (1955).
4. DOHERTY, D. G., SHAPIRA, R. and BURNETT, W. T., *J. Amer. Chem. Soc.*, **79**, 5667 (1957).
5. ELDJARN, L., PIHL, A. and SHAPIRO, B., *Proc. Intern. Conf. on the Peaceful Uses of Atomic Energy*, **11**, 335 (1956).
6. ELDJARN, L. and PIHL, A., *Progress in Radiobiology*, p. 249 Oliver & Boyd, Edinburgh (1956).
7. ELDJARN, L. and PIHL, A., *J. Biol. Chem.*, **225**, 299 (1957).
8. FRUTON, J. S. and CLARKE, H. T., *J. Biol. Chem.*, **106**, 667 (1934).
9. HAGEN, U., *Advances in Radiobiol. (Lond.)*, 187 (1957).
10. KHYM, J. X., SHAPIRA, R. and DOHERTY, D. G., *J. Amer. Chem. Soc.*, **79**, 5663 (1957).
11. KOCH, R., *Forschr. f. Geb. Röntgenstrahlen u. Nukl. Medizin*, **85**, 767 (1956).
12. LANGENDORFF, H., KOCH, R. and HAGEN, U., *Strahlentherapie*, **95**, 238 (1954).
13. ROMANO, A. H. and NICKERSON, W. J., *J. Biol. Chem.*, **208**, 409 (1954).
14. SHAPIRA, R., DOHERTY, D. G. and BURNETT, W. T., *Radiat. Res.*, **7**, 22 (1957).
15. STRAUBE, R. L. and PATT, H. H., *Proc. Soc. Exptl. Biol. Med.*, **84**, 702 (1953).
16. SUZUOKI-ZIRO, SUZUOKI-TUNEKO, *Nature*, **173**, 83 (1954).
17. TOENNIES, G. and CALLAN, T. P., *J. Biol. Chem.*, **129**, 481 (1939).

Effect of Protective Doses of *l*-Cysteine on the Level of Non-Protein Sulphydryl Groups in the Tissues of Rats Exposed to X-Rays

L. S. Isupova

THE HYPOTHESIS concerning the active participation of SH-groups in the primary biochemical reactions which take place in the exposed organism has been widely discussed in the literature.

There is a fairly large amount of material indicating the possibility of weakening the biological effect of radiations by the prophylactic administration of certain sulphur-containing, particularly sulphydryl, compounds to animals. However, the question of the accumulation of SH-groups in the tissues of animals when protective substances are given has been studied very little. Nor does the literature contain any information on the determination of non-protein sulphydryl groups in organs and tissues at different times after the exposure of animals to which large "protective" doses of SH-compounds have been given before irradiation.

In order to study the mechanism of the action of thiol prophylactics, investigations were carried out on male white rats weighing 180–220 g. The source of SH-groups introduced into the body was highly purified *l*-cysteine, which was injected intraperitoneally in doses of 100 mg/ 100 g weight (calculated on the free base) in the form of a 10 per cent solution at pH 6·9.

All the test animals were divided into groups: (a) healthy animals, (b) healthy animals to which *l*-cysteine had been given, (c) animals exposed to X-rays in doses of 700–1400 r, (d) animals exposed to X-rays with the previous injection of *l*-cysteine. Rats were decapitated at different times, and the liver and spleen were immediately removed, weighed and homogenized. In order to prevent the catalytic oxidation of SH-groups in the presence of traces of metals, sodium ethylenediaminetetraacetate was added. To precipitate the proteins a 4 per cent

55

solution of sulphosalicylic acid was added and the homogenization was repeated. The homogenates were made up to a definite volume and filtered. All the operations with the organs were carried out in ice, and the solutions used were previously cooled. To determine the quantity of SH-groups, aliquots (5 ml.) of proteinless filtrates were measured into 50 ml. conical flasks. An equal amount of acetate buffer (pH 5·4) was added, and 2 ml. of a solution of p-chloromercuribenzoic acid (0·005 N), was run into each flask. The excess of p-chloromercuribenzoic acid was back-titrated with a standard solution of l-cysteine (0·005 N). The results were expressed in milligrammes per cent of SH groups. This method gives the most reliable results as compared with other methods known from the literature.

Conditions of exposure: voltage 180 kV, current strength 15 mA, dose-rate 40 r/min; focal length 40 cm; filter 0·5 mm Cu + 1 mm Al.

As the observations showed, a total dose of 700 r for the rats under the given conditions is the minimum 100 per cent lethal dose, independent of the time of year, sex, age and weight of the animals.

In selecting the organs for study (liver and spleen), the choice was made for the following reasons:

1.   The liver is the the most important site of the metabolism of "biogenic thiols" (biosynthesis of glutathione and other SH-compounds, oxidative decomposition of l-cysteine and so on). It is moreover known from the literature[1] that it is in the liver that the maximum accumulation of SH-glutathione takes place after it has been injected into mice and rats in large doses (500 mg/100 g weight). Finally, we (Ivanov, Isupova, Yakovlev) showed that among the organs which accumulate the labelled (^{35}S) sulphur of injected l-cysteine, the liver occupies one of the leading places.

2.   The spleen was studied as one of the organs which take part in blood-forming processes and are most strongly affected by the action of radiation. In addition, the significance which is given to experiments on screening the spleen, and also on transplanting it, for prophylactic and restorative purposes in radiation sickness, was taken into account[2].

In an experiment on 30 rats it was shown that in healthy animals the range of the biological variation in the content of SH-groups for liver is 16–22 mg per cent, and for spleen 13–20 mg per cent.

In the tables the average values are given.

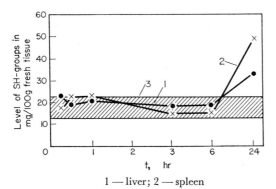

1 — liver; 2 — spleen

FIG. 1. Effect of exposure on the level of non-protein
SH-groups in the liver and spleen of rats

In Table 1 and Fig. 1 results are given which were obtained on
exposed animals into which no sources of SH-groups had been
injected. Thus, in this case it was possible to trace the effect of exposure
on the content of preformed non-protein SH-groups of the liver and
spleen at different times after the action of filtered X-rays in doses of
700 and 1400 r.

TABLE 1

RESULTS OF A STUDY OF THE EFFECT OF EXPOSURE ON THE LEVEL
OF PREFORMED NON-PROTEIN SH-GROUPS IN THE LIVER AND
SPLEEN OF RATS

(Number of animals in experiment 4–6)

Time after exposure	Level of non-protein SH-groups, mg/100 g tissue			
	Liver		Spleen	
	700 r	1400 r	700 r	1400 r
5 min	—	17·20	—	17·40
15 min	23·00	26·30	17·30	26·00
30 min	19·90	24·80	22·00	22·00
1 hr	20·96	23·00	21·50	18·50
3 hr	18·50	19·60	14·00	25·10
6 hr	19·00	—	14·50	—
24 hr	33·00	30·30	49·70	39·60

From the results in Table 1 it is seen that with a dose of 700 r the level of non-protein SH-groups in these organs in the first hours after exposure hardly changes and is within the limits of the ordinary fluctuations of the normal. In additional experiments it was shown that directly after exposure a slight reduction in the quantity of non-protein SH-groups is noted but is already levelling off after 15 min. In the literature this transient reduction is treated as a reaction in answer to irradiation of the hypophysis-suprarenal system[3]. However, 24 hr after exposure an increased content of SH-groups is found in the liver. (A similar phenomenon was noted by Rausch and Graul[5] for mouse liver.) In the spleen after 24 hr an increase in the content of non-protein SH-groups compared with normal was observed, which was still more pronounced than in the liver.

In accordance with the results of other authors[4], the experiments carried out did not show any visible change in the level of non-protein SH-groups of the liver and spleen in the early stages after exposure to X-rays in doses of 700 and 1400 r. As regards the increased content of thiol groups observed after 24 hr, this rise is evidently the result of metabolic processes changed by exposure.

As regards the spleen the interpretation of the results obtained is complicated somewhat by the fluctuations in weight of this organ in rats under normal conditions and particularly after the effect of irradiation. Twenty-four hours after exposure, the weight of the spleen decreases by about a half. However, the change in level of the SH-groups in the spleen after the action of radiation takes place in the same direction as in the liver.

To clear up the problem of the accumulation of thiol groups in the organs when protective doses of aminothiol compounds are introduced into the animals and at the same time to study the effect of exposure, experiments with l-cysteine were conducted. The results of these experiments are given in Figs. 2 and 3 and in Tables 2, 3 and 4.

In the non-exposed rats (Table 2) the content of SH-groups after the injection of l-cysteine increases by more than twice compared with the upper limit of the normal even after 5 min. This raised level is maintained for a fairly long time; for the extent of the 3 hr period of observation. After 24 hr the amount of non-protein SH-groups in the liver diminishes. but still keeps above the upper limit of the normal.

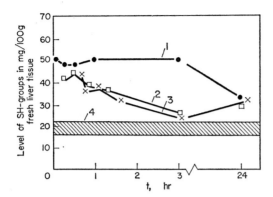

1 — non-exposed rats; 2 — rats exposed to a dose of 700 r; 3 — rats exposed to a
dose of 1400 r; 4 — biological scatter

FIG. 2. Level of non-protein SH-groups in the liver of
rats at different intervals of time after giving *l*-cysteine,
and exposure

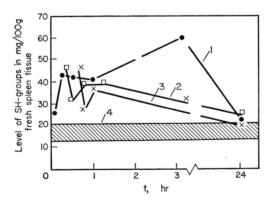

1 — non-exposed rats; 2 — rats exposed to a dose of 700 r; 3 — rats exposed to a
dose of 1400 r; 4 — biological scatter

FIG. 3. Level of non-protein SH-groups in the spleen
of rats at different intervals of time after giving
l-cysteine, and exposure

TABLE 2

RESULTS OF A STUDY OF THE LEVEL OF NON-PROTEIN SH-GROUPS IN
THE LIVER AND SPLEEN OF RATS AFTER GIVING *l*-CYSTEINE

(Number of animals in experiment, 4–6)

Time after injection of compound	Level of non-protein SH-groups, mg/100 g weight of tissue	
	Liver	Spleen
5 min	50·10	25·50
15 min	48·20	43·50
30 min	48·00	42·50
1 hr	50·70	40·20
3 hr	49·00	60·00
24 hr	31·70	21·00
72 hr	28·60	22·60

TABLE 3

RESULTS OF A STUDY OF THE LEVEL OF NON-PROTEIN SH-GROUPS IN
THE LIVER AND SPLEEN OF RATS AFTER GIVING *l*-CYSTEINE AND
EXPOSING TO X-RAYS (700 R)

(Number of animals in experiment, 4–6, duration of exposure, 17 min)

Time from injection of compound to killing	Time after exposure	Level of non-protein SH-groups, mg/100 g weight of raw tissue	
		Liver	Spleen
22 min	5 min	41·10	45·50
32 min	15 min	43·60	31·00
47 min	30 min	38·40	37·00
77 min	1 hr	35·40	38·50
3·3 hr	3 hr	25·20	—
24·3 hr	24 hr	29·00	22·40

Quite another picture is observed in rats which after the injection of *l*-cysteine were subjected to irradiation (Fig. 2; Tables 3, 4). In comparison with healthy animals, to which *l*-cysteine was also given, in these animals the level of non-protein SH-groups in the liver falls during the first minutes and hours after exposure. It can be imagined that a decrease in SH-groups took place during the exposure itself.

Three hours after exposure, the content of SH-groups in the liver reached the upper limit of the normal, i.e. it was reduced by more than 50 per cent from the original value. Increasing the exposure dose two-fold had no substantial effect on the character of the change in level of the titratable SH-groups of the liver.

For the spleen a similar rule was noted (Fig. 3; Tables 3, 4). The decrease in the thiol groups was more clearly pronounced at a dose of 1400 r than at a dose of 700 r.

Thus, as a result of the experiments it was found that in the liver and spleen of rats into which no SH-compounds had been introduced from outside, the content of sulphydryl groups in the first hours after exposure did not exceed the limits of biological variation. After 24 hr a certain rise in the level of SH-groups in these organs was noted.

It was also shown that the curves of the decrease in content of SH-groups with time, in rats that had received *l*-cysteine, differ considerably for non-exposed animals and for those subjected to the action of radiation.

In the injected animals a more marked decrease in SH-groups in the liver and spleen was noted soon after exposure.

Doubling the minimum lethal exposure dose at the same dose-rate had little effect on the character of the decrease in SH-groups in the liver and spleen of rats that had received large doses of *l*-cysteine.

On comparing the results on the change in content of SH-groups in the liver and spleen at different times after giving a protective substance to rats, with and without their exposure, it may be suggested that the primary conversions of *l*-cysteine introduced into the body are not the same for the non-exposed and the exposed animals. The form of combination taken up by the SH-groups of injected *l*-cysteine is not yet known. This may later be elucidated.

It may be imagined that the *l*-cysteine introduced into the body is first concentrated in the liver and in a normal animal gradually enters into biochemical reactions, taking part, for example, in the biosynthesis of SH-glutathione, the SH-form of coenzyme A and other thiol compounds. In the exposed body the participation of *l*-cysteine in biosynthetic reactions may be disturbed; as a result of this the excess *l*-cysteine may undergo rapid radiochemical decomposition by oxidation, thus protecting endogenous thiol groups from the harmful action of radiation.

TABLE 4

RESULTS OF A STUDY OF THE LEVEL OF NON-PROTEIN SH-GROUPS IN THE LIVER AND SPLEEN OF RATS AFTER GIVING *l*-CYSTEINE AND EXPOSING TO X-RAYS (1400 R)

(Number of animals in experiment, 4–6, duration of exposure, 35 min)

Time from injection of compound to killing	Time after exposure	Level of non-protein SH-groups, mg/100 g weight fresh tissue	
		Liver	Spleen
40 min	5 min	43·0	45·20
50 min	15 min	37·0	27·00
65 min	30 min	37·4	36·50
3·6 hr	3 hr	31·0	32·20
6·6 hr	6 hr	23·1	32·20
24·6 hr	24 hr	30·0	19·00

The possibility is not excluded that the rate of oxidation of the excess SH-compounds is increased as a result of the physiological condition of the animals which has been changed by exposure, with a general intensification of metabolic processes during exposure and at times soon after it.

The final elucidation of the role of sulphydryl groups in problems of protection of the body from radioactive emissions requires further investigation.

REFERENCES

1. CRONKITE, E. P., CHAPMAN, W. H. and BRECHER, G., *Proc. Soc. Exptl. Biol. Med.*, **76**, 456 (1951).
2. JACOBSON, L. O., *Amer. J. Roentgenol.*, **72**, 4, 543 (1954).
3. KOCH, R., *Forschr. f. Geb. Röntgenstrahlen u. Nukl. Medizin*, **85**, H.6, 787 (1956).
4. PETERSON, R. D., BEATTY, C. H. and WEST, E. S., *Proc. Soc. Exptl. Biol. Med.*, **77**, 747 (1951).
5. RAUSCH, L. and GRAUL, E. H., *Strahlentherapie*, **94**, 539 (1954).

Effect of Protective Substances on the Protein SH-Groups in Organs and Tissues of Healthy and Exposed Animals

V. G. Yakovlev and L. S. Isupova

In the previous survey it was shown that there was a relationship between the prophylactic effect and the concentration of low molecular-weight thiolic substances in the organs and tissues of rats to which protective sulphur-containing compounds had been administered before exposure. In continuation of the investigation, the effects of exposure and protective substances on the content of protein SH-groups of rat tissues have been studied. In addition known data, concerning the dynamic equilibrium between the SH-groups of protein and the low molecular-weight thiols and disulphides contained in the tissue fluids in physiological concentrations, have been taken into account. It was to be expected that after the introduction of comparatively large amounts of protective (very reactive) compounds into the body, they would enter into chemical reaction with the different functional groups of the biomolecules, including protein SH-groups.

In the radiobiological literature the hypothesis has repeatedly been put forward that the chain of pathological processes arising in the exposed organism is connected with the action of radiation on the sensitive SH-groups of proteins and, in particular, enzymes[1, 2], such action being either direct, or through the products of the radiolysis of water. However, in spite of the considerable number of investigations made in this direction, the problem of the effect *in vivo* of minimum lethal doses of radiation on the proteins and SH-enzymes is still not sufficiently clarified. Thus, data concerning whether the level of protein SH-groups of the organs and tissues changes immediately after exposure of the organism are contradictory. As regards the action of radiation on enzyme systems for the same species of animals (e.g. rats), according to the literature data, the activity of some SH-enzymes decreases immediately after exposure, and the activity of others increases under the action of radiation *in vivo*[3-5]. Regarding the data on the reduction in the content of protein[18] and non-protein SH-groups of tissues at comparatively late times after exposure ("radiation after-effect"), it is possible that this occurs as a result of the inactivation of the SH-groups by organic peroxides formed in the exposed body as a result of chain reactions[19]. Thus, the problem of the active participation of protein SH-groups and non-protein SH-compounds in the primary radiation-biochemical processes requires further study. This is particularly necessary in connection with the recent repeated pronouncements of Norwegian investigators (Eldjarn, Pihl *et al.*) concerning the protective action of compounds containing sulphur[6-11]. The essence of these hypotheses is that substances of the cysteamine-cystamine group react extensively *in vivo* with disulphide and thiol groupings in protein molecules, with the formation of short-lived mixed disulphides, and as a result of this, the biogenic sulphur-containing groups become more stable to the action of radiation. There is, however, no direct evidence indicating a relationship of the protective effect with the formation of temporary linkages of the protective substances with the SH- or —S—S-groups of protein molecules. In connection with this the problem of the state of the protein SH-groups under conditions of exposure and of chemical protection required further experimental study.

The experiments were carried out on male white rats weighing 180–200 g. Free *l*-cysteine and cysteamine hydrobromide were used as protective substances, which were studied from the point of view of their effect on the condition of the protein SH-groups. The content of SH-groups of the liver and testis tissue was determined. The experimental animals were divided into five groups.

The first group consisted of healthy rats to which *l*-cysteine (10 per cent solution, pH 6·9) was given intraperitoneally in a dose of 100 mg/100 g weight, 30 min before death.

The second group was of healthy rats to which cysteamine HBr was given, calculated at 21 mg/100 g body weight, 30 min before death.

The third group was of rats which were exposed to X-rays at a dose of 650 r, and were killed immediately*.

The fourth group was of rats to which *l*-cysteine was given in the dose indicated above, 12 min before exposure, and which were then exposed and killed immediately after exposure (30 min after introduction of the substance).

The fifth group consisted of animals that had received cysteamine–HBr with subsequent exposure. The solution was injected 5 min before exposure, and they were killed 7 min after exposure. For each series of experimental rats a group of healthy control animals was studied (i.e. without the injection of the substances and without exposure). In additional experiments with *d,l*-cysteine labelled with ^{35}S, the radioactivity of the liver and testis tissues was determined 30 min after the intraperitoneal injection of the indicated doses (6000 counts/min/g weight) of labelled cysteine in a mixture with protective doses of ordinary cysteine.

The level of ordinary protein SH-groups was determined in homogenized solutions of tissues, prepared by means of denaturing agents. It was, however, thought that the results of the determination of the total protein SH-groups could give a more definite answer to the question of the action of penetrating radiation on these very important functional groups of the protein molecules. It should be considered that the differences between the so-called unhindered or freely reacting and other categories of protein SH-groups, sluggishly reacting and masked, are very nominal. These differences may be made apparent as a result of the use of chemical reagents which differ in oxidative action or degree of affinity for the sulphur of the SH-groups. The reactivity of the latter in turn greatly depends on the presence and character of the different substituents, their arrangement in the molecule, "ionic environment" and other conditions. Considerable importance attaches to the geometrical configuration of the protein molecules and the steric hindrance connected with it which affects the reactivity of the SH-groups. However, during the action of penetrating radiation on proteins steric hindrance can hardly be significant, so

---

* The exposure was carried out on a RUM-3 apparatus, voltage 180 kV, current strength 15 mA, filter 0·5 mm Cu + 1 mm Al, focal length 40 cm, dose-rate 34–35 r/min, time of exposure 18 min. With these conditions, in the control experiments 97–100 per cent of the rats died from radiation disease 10–15 days after exposure.

that the SH-group which is "hidden" in the chemical sense may be subjected to the direct or indirect action of penetrating radiation. As regards the various artificially obtained fractions of proteins (for example, "water-soluble" and so on), there is always a danger of loss of SH-groups as a result of oxidation or other causes during various operations connected with the preparation of these fractions. The method of preparing the protein "solutions" must answer the requirements of speed of preparation and must guarantee maximal resistance of the SH-groups to external effects during the analytical determination of these groups.

By using a combination of the denaturing agents urea and guanidine hydrochloride, we obtained homogeneous solutions of liver and testis tissues. Such solutions could not successfully be prepared for the spleen, brain and certain other organs.

Evaluating the possible difficulties in developing the method, we took into account the observations of Pasynskii and Chernyak[13], who showed that "the oxidizability of the SH-groups in sulphydryl compounds under the influence of the oxygen of the air is considerably accelerated in the presence of high concentrations of urea and similar substances". The acceleration of oxidation, in their opinion, depends on the effect of NH_2-groups ". . . at the same time the capacity of an NH_2-group to be a proton acceptor is of significance . . .". This circumstance urged us to study the rate of oxidation by the oxygen of the air of a number of low molecular-weight thiol compounds and proteins in aqueous solutions under different conditions. In experiments with l-cysteine, cysteamine HCl and other thiols, and also with proteins it was shown that urea actually increased the rate of oxidation of the SH-groups. However, this increase in oxidation most probably depends on contamination of the reagent (urea) with catalysts for the oxidation of the SH-groups (evidently, traces of metals). Purifications of the preparation of urea by recrystallization gradually reduced its "oxidizing" action, and the addition of ethylenediaminetetraacetic acid removed it completely. As regards guanidine hydrochloride, in the chemically pure state it not only did not accelerate the oxidation of the SH-groups, but on the contrary it very much increased their resistance to the oxidative effect of the oxygen of the air.

In the literature known to us we did not find any indications of the existence of a method of treating proteins, that would guarantee the complete liberation of all "hidden" SH-groups. This is bound up with certain difficulties in method.

After numerous trial experiments we found the best ratios of urea (1·33 mmole/ml.), guanidine HCl (4·4 mmole/ml.), the disodium salt of EDTA (1 mg/ml.) and tissues (40 mg/ml.) in the total solution. With these conditions the maximum quantity of tissue SH-groups was determined. A considerable increase in the concentration of "detergents", as also a number of other variations (heating in anaerobic conditions, change of pH and so on), did not lead to an increase in the titratable SH-groups. During the incubation of tissue solutions (20°, 37°) in open vessels a reduction in SH-groups was marked only after 20 hr. When reduced glutathione, l-cysteine, its n-propyl ester HCl, cysteamine HCl and a number of other thiolic substances were added to these solutions their SH-groups were determined quantitatively as a total with those of the tissue. Addition of the S—S-forms of the compounds enumerated did not affect the level of SH-groups either immediately after the introduction of disulphides into the solution, or on incubation for 10–20 hr at 20°. It is well known from the literature that the disulphide linkages of the protein insulin are not reduced on denaturing to the SH-form. The constancy in the content of SH-groups of the tissue solutions studied (containing denaturing agents) when they are in contact with the air for a comparatively long time indicates that ethylenediaminetetra-acetic acid does not interfere. Some misgivings on this point had been based on certain literature results to the effect that the oxidation of the SH-groups of tissue glutathione by the oxygen of the air is accelerated in the presence of EDTA and ascorbic acid[14]. It may be suggested that under our experimental conditions this effect did not occur at all.

In its final form the method consisted of the following: rats were decapitated, and an organ (for example, the liver) was removed, weighed and quickly ground up in a mortar (in ice). A 1 g portion of tissue was homogenized with 2 g dry, chemically pure coarsely crystalline urea and 25 mg EDTA, part of a solution of a second detergent (10·5 g guanidine hydrochloride in twice-distilled water) was added, the homogenization was repeated and the mass was washed into a measuring vessel with the rest of the guanidine solution. The volume was made up to 25 ml. with water, mixed and 3–4 samples of

5 ml. each, in small flasks, were taken from the clear homogeneous solution. Five millilitres 0·2 N acetate buffer (pH 5·3) and a measured quantity (2 ml.) of mercury reagent (an exactly 0·005 M solution of *p*-chloromercuribenzoic acid in a 0·01 M solution of KOH) were added to each of the samples. After 10 min the excess of unreacted reagent was back-titrated with a standard 0·005 M solution of cysteine ethyl ester hydrochloride. The end point was found by means of the dry nitroprusside reagent as described by McDonnel[15]. Control experiments, in which distilled water was taken instead of the liquid to be analysed, were carried out in parallel. The results were expressed in milligrammes SH-groups per 100 g crude tissue. By this method the total amount of tissue SH-groups was determined. In parallel experiments the content of SH-groups of the proteinless filtrates of the same tissue was determined (the proteins were precipitated with 4 per cent sulphosalicylic acid). The total amount of protein SH-groups was calculated from the difference between the total tissue and non-protein SH-groups. The results obtained were subjected to statistical treatment.

The total amount of protein SH-groups in the liver tissue of healthy rats varied from 72 to 75 mg per cent. The quantity of non-protein SH-groups in liver was less constant and varied from 16 to 25 mg per cent, depending on the time of year, diet and a number of other conditions. As regards testis tissues, in them the content of non-protein SH-groups (17–18 mg per cent) approximated to their content in liver, whereas the amount of protein groups was 2–2·5 times lower.

In Table 1 the effect of exposure on the total amount of protein SH-groups in the liver and testis tissues of white rats, killed immediately after exposure, is given.

TABLE 1

LEVEL OF PROTEIN SH-GROUPS IN LIVER AND TESTIS TISSUES OF WHITE RATS (mg %)

Organ	Before exposure	After exposure	Change, % of normal
Liver	71·7 ± 1·83	66·1 ± 2·3	92·2
Testis	31·9 ± 1·87	23·1 ± 1·35	72·4

As seen from Table 1, the total amounts of protein sulphydryl groups in the tissues investigated are thereby decreased.

This reduction is indicated more clearly for testes (by 28 per cent), the tissue of which is known to be more sensitive to the harmful action of penetrating radiation.

In Table 2, results are given on the effect of protective substances (*l*-cysteine and cysteamine hydrobromide) on the level of non-protein and total protein SH-groups in the liver and testes of unexposed rats killed 30 min after the administration of these substances.

TABLE 2

EFFECT OF *l*-CYSTEINE AND CYSTEAMINE (HBR SALT) ON THE LEVEL OF NON-PROTEIN AND TOTAL PROTEIN SH-GROUPS IN THE LIVER AND TESTIS TISSUES OF NON-EXPOSED RATS

Organ	Compound	Dose of substance, mg/100 g	SH-groups introduced, mg/100 g	Level of non-protein SH-groups, % of normal	Level of total protein SH-groups, % of normal
Liver	*l*-Cysteine (free)	100	27·3	203·0	83·4
Liver	Cysteamine (HBr salt)	21	4·4	114·6	105·3
Testis	*l*-Cysteine (free)	100	27·3	128·6	91·5
Testis	Cysteamine (HBr salt)	21	4·4	110·8	88·1

Note—The substances were given intraperitoneally. In the determinations the variations did not exceed the values given in Table 1.

It is seen from Table 2 that there are substantial differences in the condition of the SH-groups depending not only on peculiarities of structure and chemical properties (for example, oxidizability) of the protective compounds, but evidently also on the nature of the proteins in the tissues studied.

The differences in action of various thiolic substances on protein are detected on examining the results for liver tissue. Thus, after the administration of *l*-cysteine the total amount of protein SH-groups is considerably decreased (by about 17 per cent). This curious fact may be an indirect indication that certain protein SH-groups enter partially and probably reversibly into reaction with the cysteine introduced, forming a mixed disulphide according to the following equation:

$$\text{Protein—SH} + \text{HS—R} - 2e \rightleftarrows \text{Protein—S—S—R} + 2H^+$$

where R is the residue of the cysteine molecule.

In the literature[16] there is an indication of a similar reaction *in vitro* between SH-peptides and cysteine. It is not beyond the bounds of possibility that the reaction of SH-proteins with SH-compounds is one of the modes of detoxication of an excess quantity of these compounds.

In contrast to *l*-cysteine, cysteamine not only does not reduce but, on the contrary, slightly raises the level of protein SH-groups in liver. It is possible that cysteamine is actually, as Norwegian research workers suggest[10], capable of reducing the disulphide bonds in protein molecules with the intermediate formation of mixed disulphides according to the following equation:

A. | Protein |‒S‖S   + R—SH ⇄ | Protein |‒SH / ‒S—S—R

B. | Protein |‒SH / ‒S—S—R   + R—SH ⇄ | Protein |‒SH / ‒SH   + R—S—S—R

where R is a cysteamine residue. Disulphide prophylactics (R—S—S—R), in the opinion of the authors referred to, react according to reaction B from right to left. In testis tissue, however, both *l*-cysteine and cysteamine reduce the content of protein SH-groups, i.e. they act as though in one direction. It is as yet difficult to explain such differences in the properties of proteins of different tissues. It can only be suggested that there are considerably fewer protein —S—S- bonds in testis tissue than in liver tissue.

In Table 2, results are also given for non-protein SH-groups. As seen from the table, after the administration of cysteine the amount of non-protein SH-groups in liver is doubled. Cysteamine penetrates into liver tissue just as much as *l*-cysteine. The comparatively small increase in this case (about 15 per cent) in the level of non-protein SH-groups is quite normal if it is taken into account that in terms of molecular proportions the amount of aminothiol introduced into the body was 6·2 times less in comparison with *l*-cysteine. It is interesting to note that the extent of penetration of cysteine SH-groups into testis tissue is considerably smaller than into liver tissue. For cysteamine this difference is considerably less pronounced. By experiments with cysteine labelled with ^{35}S it could be shown that 30 min after the introduction of this amino-acid into the body of the rat, the total radioactivity of the testis tissue was several times smaller than the radioactivity of the liver tissue.

In Table 3, results are given on the change in content of non-protein and protein SH-groups of tissues of rats to which protective substances have been given and which have afterwards been exposed to X-rays in lethal doses (650 r). The "protected" exposed animals were killed 30 min after injection of the chemical compounds.

TABLE 3

EFFECT OF *l*-CYSTEINE AND CYSTEAMINE (HBr SALT) ON THE LEVEL OF NON-PROTEIN AND TOTAL PROTEIN SH-GROUPS IN THE LIVER AND TESTIS TISSUES OF EXPOSED RATS

Organ	Compound	Dose of substance, mg/100 g	SH-groups introduced, mg/100 g	Level of non-protein SH-groups, % of normal	Level of total protein SH-groups, % of normal
Liver	*l*-Cysteine (free)	100	27·3	214·3	101·7
Liver	Cysteamine (HBr salt)	21	4·4	109·3	100·7
Testis	Cysteamine (HBr salt)	21	4·4	—	76·2

Note—The substances were given intraperitoneally. Rats were exposed for 12 min and killed immediately after exposure.

It is seen from Table 3 that the effect of protective substances on the condition of the protein SH-groups in the exposed animals is of a different character from that in animals which were not exposed. Thus, the total amount of protein SH-groups in the liver after the introduction of both substances hardly differed from the amount of these groups in the liver of healthy animals. However, in the case of testes and the administration of cysteamine the content of protein SH-groups was decreased only slightly less than in exposed unprotected rats.

The absence of any change in the content of protein SH-groups in the liver tissue in "protected" and exposed rats is difficult to explain. If the decrease in protein SH-groups under the influence of radiation is connected with their oxidation, then in the presence of easily oxidized protective substances this process of oxidation of protein SH-groups may be diminished. On the other hand, the possibility is not excluded that under the action of radiation decomposition of the mixed disulphides takes place by those mechanisms, the existence of which Pihl and Eldjarn consider probable[10]. They suggest that the free radicals may attack either of the —S—S-atoms and, if the sulphur of the cysteamine is oxidized by the free radical, the SH-group of the protein will be restored in its original condition. The same assumption was made with regard to direct hits. It is known that in an ionized organic molecule the charge may migrate along the carbon chain until a bond breaks[7]. If the rupture should occur at the sensitive mixed disulphide bond, this would represent a (partially) non-destructive mode of dissipation of energy.

In this paper it has been shown that the total content of protein SH-groups changes both under the action of radiation and under the action of protective substances. The changes for different substances and for different proteins in the animal body are not well defined. The nature of these phenomena has not yet been explained. It may, however, be considered that the inactivitation of the protein SH-groups both immediately after exposure and during the development of radiation disease is of considerable importance in the mechanism of the action of radiation on the animal body. The possibility of the rupture of inter-molecular linkages formed with the participation of SH-groups is not excluded, which may lead to the disorganization of biological structures. In addition, the possible generation, in protein molecules, of groupings containing sulphur of a high degree of oxidation, would lead to the disturbance of metabolic processes.

In the mechanism of the protective action of aminothiols and SH-aminoacids an important part is played, evidently, by the reaction of these compounds with the reactive parts of the proteins, purine and pyrimidine bases and other biomolecules. The problem of the chemical reaction of protective substances with the different functional groups of the tissue molecules requires further study. It may give important information about the mechanism of the action of radiation on

D

biological systems and suggest new methods for the prophylaxis of radiation sickness.

REFERENCES

1. BARRON, E. S. G., The effect of ionizing radiations on some systems of biological importance, in *Symposium on Radiobiology* (J. J. Nickson, ed.) p. 216 (1952).
2. BARRON, E. S. G., The *in vitro* effects of radiation on molecules of biological importance, in *Nuclear Science Series* Rept. No. 17, p. 30, Washington (1954).
3. ROTH, J. S., EICHEL, H. J., WASE, A., ALPER, C. and BOYD, M. J., *Arch. Biochim. Biophys.*, **44**, 95 (1953).
4. DUBOIS, K. P. and PETERSEN, D. F., *Amer. J. Physiol.*, **176**, 282 (1954).
5. ORD, M. and STOCKEN, L. A., *Brit. J. Radiol.*, **28**, 279 (1955).
6. ELDJARN, L., PIHL, A. and SHAPIRO, B., *Proc. Intern. Conf. on the Peaceful Uses of Atomic Energy*, **11**, 335 (1956).
7. SHAPIRO, B. and ELDJARN, L., *Radiat. Res.*, **3**, 255, 393 (1955).
8. ELDJARN, L. and PIHL, A., *J. Biol. Chem.*, **223**, 341 (1956).
9. ELDJARN, L. and PIHL, A., *Progress in Radiobiology*, p. 239, Oliver & Boyd, Edinburgh (1956).
10. PIHL, A. and ELDJARN, L., *Advances in Radiobiology*, Oliver & Boyd, Edinburgh (1957).
11. ELDJARN, L. and PIHL, A., *J. Biol. Chem.*, **225**, 200 (1957).
12. BARRON, E. S. G., *Advances in Enzymology*, **11**, 201 (1951).
13. PASYNSKII, A. G. and CHERNYAK, R. S., *Biokhimiya*, **17**, 198 (1952).
14. PIRIE, A. V. and HEYNINGEN, R., *Nature*, **173**, 873 (1954).
15. MACDONNEL, L. R., SILVA, R. B. and FEENEY, R. E., *Arch. Biochim. Biophys.*, **32**, 288 (1951).
16. LIVERMORE, A. H. and MUECKE, E. C., *Nature*, **173**, 265 (1954).
17. WALLENSTEIN, M., WAHRHAFTIG, A. L., ROSENSTOCK, H. and EYRING, H., *Symposium on Radiobiology*, p. 70, Wiley, New York (1952).

Synthesis and Testing of the Protective Action of a Number of Sulphur-containing Compounds and Derivatives of Coumarin

V. G. YAKOVLEV and V. S. MASHTAKOV

IN THE previous communications, data were given on the testing of the protective properties of aminothiol compounds. The task of the present investigation was to look for new series of chemical compounds, both sulphur-containing and sulphur-free, and possessing protective action against penetrating radiations.

For this purpose the following were synthesized and tested on animals:

(a) some organic nitrogen thio-compounds;

(b) polythionic acids;

(c) derivatives of coumarin.

The experiments on protection were carried out on white rats weighing 180–240 g, which after injection of the chemical compounds were exposed to X-rays in lethal doses (600–650 r).

For the first series of experiments compounds with $C = S$ groupings were synthesized (for example, dithiocarboxylic acids of the general formula $R - \underset{\underset{S}{\|}}{C} - SH$), and also substances in which the mutual

positioning of the $C = S$ and NH_2-groups makes possible the formation of tautomeric SH-forms according to the equation:

$$-\underset{\underset{S}{\|}}{C} - NH_2 \quad \rightleftarrows \quad -\underset{\underset{SH}{|}}{C} = NH$$

The chemical characteristics of six substances of this group, which were tested on animals, are briefly given below.

1. The thioanhydride of dithiocarbamic acid (thiuram sulphide)

$$H_2N - \underset{\underset{S}{\|}}{C} - S - \underset{\underset{S}{\|}}{C} - NH_2 \quad \rightleftarrows \quad HN = \underset{\underset{SH}{|}}{C} - S - \underset{\underset{SH}{|}}{C} = NH$$

was obtained as the diammonium salt in fairly good yield by the method of Hlasiwetz and Kachler, described in Beilstein[10]. The substance is easily oxidized by $FeCl_3$ to thiuram disulphide.

2. Potassium trithioallophanate monohydrate

$$H_2N-\underset{\underset{S}{\|}}{C}-NH-\underset{\underset{S}{\|}}{C}-SK\cdot H_2O$$

was synthesized in almost quantitative yield from thiourea and CS_2 by the method described by Rosenheim and co-workers[23], and was identified in the form of the ethyl ester (m.p. 173°, according to the literature 174°).

3. Sodium azidodithioformate dihydrate

$$N_2:N-\underset{\underset{S}{\|}}{C}-SNa\cdot 2H_2O$$

was prepared by Sommer[26] from sodium azide and carbon disulphide with a yield of pure product of up to 60 per cent of the theoretical.

The synthesis of the substance is simple. In the crystalline form with two molecules of H_2O this compound is fairly stable. In the anhydrous condition it decomposes with explosion when rubbed or heated and therefore requires care in handling. In well closed jars and at a temperature below 10° the crystalline hydrate keeps for a long time without decomposition. Azidodithioformic acid is easily oxidised to the S—S-form. The Li and K salts cannot be isolated because of their instability.

4. A complex derivative of dithiocarbamic acid, a condensation product of glycine ethyl ester with CS_2 having (according to E. Fischer[13]) the structure:

$$C_2H_5\cdot COO\cdot CH_2-NH-\underset{\underset{S}{\|}}{C}-SH\cdot NH_2-CH_2CO\cdot O\cdot C_2H_5$$

This substance, being salt-like, is readily soluble in water but on keeping even in the cold gradually decomposes, acquiring a rose colour.

5. 2-Amidinothiourea was synthesized by the method of Rathke[22] and Bamberger[7] from dicyandiamidine sulphate and H_2S

$$H_2N-\underset{\underset{NH}{\|}}{C}-NH-\underset{\underset{S}{\|}}{C}-NH_2$$

and was separated by means of the sparingly soluble oxalate. The free base is stable on storage. Its aqueous solutions have an alkaline reaction and for experiments on animals they had to be neutralized with HCl.

6. S-Acetylthiourea hydrochloride was obtained by Dixon[9] in

$$HN\!=\!C\!-\!S\!-\!CO\!-\!CH_3$$
$$\underset{NH_2\cdot HCl}{|}$$

quantitative yield by treating thiourea with acetyl chloride in acetone solution. The crystalline salt is easily soluble in water, but the solution has an acid reaction, and to bring the pH to physiological values neutralization with alkali is required. However, the amount of alkali used exceeds that required in theory for combination with the HCl. This indicates the possibility of cleavage of the acetyl group. Besides the compounds enumerated, potassium dithiocarbazate[27] dithio-oxamide[8] and mercaptosulphobenzothiazole[2] were synthesized, but all these compounds proved to be toxic and could not be tested as prophylactics.

In Table 1 results are given on the testing of the protective action of the thio-compounds synthesized.

In testing substances 1–6 and the polythionates the following conditions for exposure were taken: voltage 180 kV, current strength 15 mA, radiation dose 650 r, dose-rate 40 r/min, focal length 40 cm, filter 0·55 mm, Cu + 1 mm Al.

For each experiment 18–20 rats were taken. Solutions of the substances were injected intraperitoneally at 1 ml/100 g weight 10 min before exposure. Solutions of 2-amidinothiourea and of S-acetylisothiourea were brought to pH 6·9–7·1 by means of HCl and NaOH respectively.

As seen from Table 1, 2-amidinothiourea (hydrochloride) possessed considerable protective action, as also did sodium azidodithioformate dihydrate (in a dose of 70 mg/100 g).

When the latter substance was used, the reduction of loss in weight of protected, exposed rats was considerably less than when other protective compounds (*l*-cysteine, its esters, etc.) were used. In addition, supplementary experiments showed that an increase in the dose of this substance to 100 mg/100 g caused the death of 30–50 per

TABLE 1

RESULTS OF A STUDY OF THE PROTECTIVE ACTION OF ORGANIC NITROGEN
THIO-COMPOUNDS ON RATS EXPOSED TO X-RAYS

Compound	Weight of rats, g	Dose of substance, mg/100 g	Survival rate of animals at 30 days, %	Change of weight by 30 days, % of original	Average duration of life of animals, days
Thiuramsulphide, diammonium salt	200–220	50	25	115	12
K trithioallophanate–H_2O	180–220	25	0	—	11
Na azidodithioformate–$2H_2O$	205–230	70	50	94	12
Condensation product of glycine ethyl ester with CS_2	200	50	0	—	13
2-Amidinothiourea	185	70	40	120	13
S-Acetylisothiourea–HCl	200	50	0	—	8
Control, uninjected	200–220	—	0	—	10

cent of the rats immediately after exposure, but a reduction in the dose to 50 mg/100 g led to loss of protective action.

Sodium azidodithioformate when injected in amounts of 70 mg/100 g somewhat inhibited the growth of the rats, whereas doses of 40–50 mg/100 g scarcely caused this effect. It is possible that the protective properties of this compound are in some way connected with the inhibition of mitotic processes. Thirty to forty minutes after the intraperitoneal injection of sodium azidodithioformate, the rats were observed to be greatly affected.

In spite of the cheapness of the starting materials for preparing the sodium salt and the simplicity of the synthetic method, its use is hardly possible because of toxicity.

The small though clearly pronounced protective action of 2-amidino-thiourea is of interest in that it indicates the importance of the guanidine grouping. This may be connected with the recently published results[24, 15] on the role of mercaptoalkylguanidines as possible inter-mediate products in the transformation of radioprotective isothio-uronium compounds in the body of mammals.

In a second series of experiments the sodium salts of the polythionic acids di-, tri-, tetra- and pentathionic were synthesized and tested. During the testing of the protective action of these compounds some data emerged concerning the metabolism of cysteine.

The fate of this aminoacid and the nature of the intermediate products have been explained fairly fully. There remains the question of the fate of the hydrogen sulphide that is split off from *l*-cysteine by tissue desulphydrases[14]*.

There are indications[25, 17] that the metabolism of hydrogen sulphide in the tissues (in experiments *in vitro*) is accompanied by the formation of polythionic acids which may thus appear as intermediate products in the chain of transformations of the sulphur of hydrogen sulphide in the body. This stimulated us to study the possible protective properties of polythionic acids. Sodium polythionates were synthesized in chemically pure form by methods described in the literature (the variations in the determination of sulphur were ± 0.2 per cent of the theory). Sodium dithionate was obtained by the Gay–Lussac method[1] in the form of a crystalline dihydrate ($Na_2S_2O_6 \cdot 2H_2O$). Sodium trithionate was synthesized, starting from sodium thiosulphate by a method described by Abbeg[4]. The substance crystallized with 17 molecules of water ($Na_2S_3O_6 \cdot 17H_2O$). Sodium tetrathionate ($Na_2S_4O_6 \cdot 2H_2O$) was obtained by the reaction of $Na_2S_2O_3$ with elementary iodine[4], and sodium pentathionate ($Na_2S_5O_6$) was obtained by Raschig's method[4].

The distinguishing property of polythionates (with the exception of sodium dithionate) is their capacity in aqueous solutions for gradually liberating sulphur in the colloidal condition, especially with changes of pH.

The polythionates obtained were tested on rats as protective substances. In preliminary experiments the maximum doses tolerated were found.

In Table 2, results of the tests are given. From a comparison of doses of the substances used, it is seen that the toxicity of thionates increases sharply with the increase in the number of sulphur atoms in the chain.

---

* It should not be forgotten that in experiments on protection large "non-physiological" quantities of *l*-cysteine are introduced into the body and its metabolic relationships may be other than normal.

TABLE 2

RESULTS OF A STUDY OF THE PROTECTIVE ACTION OF SODIUM POLYTHIONATES
ON RATS EXPOSED TO X-RAYS; DOSE 650 r

Compound	Weight of rats, g	Route of administration	Dose of substance, mg/100 g	Survival rate of animals at 30 days, %	Change of weight by 30 days, % of original	Average duration of life of rats that died, days
Sodium dithionate	220–240	Intra-peritoneal	300	25·0	103	17
Sodium dithionate	200–220	Oral	500	0·0	—	9
Sodium trithionate	190–220	Intra-peritoneal	144	0·0	—	7
Sodium trithionate	190–220	Oral	334	0·0	—	14
Sodium tetrathionate	210–220	Intra-peritoneal	30	0·0	—	12
Sodium tetrathionate	220	Intra-peritoneal	40	20·0	106	9
Sodium pentathionate	230–240	Intra-peritoneal	30	20·0	108	10
Sodium pentathionate	220	Intra-peritoneal	40	20·0	107	9
Control	210–220	—	—	0·0	—	9

Note—In each experiment there were 15–20 rats; intraperitoneally the substances were given 10 min before exposure, and orally 30 min before exposure. In the column "Dose of substance" the figures show the amounts of substances calculated on the anhydrous salts.

As regards protective action, a small effect (20–25 per cent) was found for the dithionate and the tetra- and pentathionates.

The mechanism of the protective action of polythionates is obscure, the more so as the dithionate (in contrast to the higher homologues) is a compound with pronounced oxidizing properties[6]. In supplementary experiments attempts were made to synthesize β-mercaptoethylamine dithionate, but more than 80 per cent of the aminothiol was oxidized at the SH-group. Cystamine dithionate was obtained in good yield by simple neutralization of the free base with dithionic acid in dilute solution. The protective effect of this salt did not exceed 20 per cent. It is also seen from Table 2 that when even considerable doses of sodium di- and trithionates were given perorally no protective action was observed.

In the following series of experiments coumarin and its sulphur-containing analogues were synthesized and tested:

coumarin   thiocoumarin   dithiocoumarin

and also derivatives of coumarin with different substituents in the nucleus, including thiol and isothiouronium groupings (Table 3).

It was found from the literature that certain hydroxy-derivatives of coumarin are antioxidants for certain oils[28]. It could be suggested that these substances might inhibit the oxidation reactions which develop after exposure in lipid systems. The coumarin derivatives were synthesized by methods described in the literature, separated in the pure form and identified by melting point.

The testing of a number of coumarin derivatives proved difficult owing to their poor solubility. Compounds even with two hydroxyl groups were sparingly soluble in water and could not be used for injections in the form of aqueous solutions. Accordingly, the compounds from No. 2 to No. 12 were introduced into the rats' stomachs in the form of suspensions in water or aqueous glycerol. No protective action was detected for any of these compounds even with the most varied doses and intervals of time between the introduction of the solutions and the exposure of the animals. In the subsequent experiments coumarin, thiocoumarin and dithiocoumarin were dissolved with slight heating in oil (a mixture of equal volumes of cod liver oil and peach oil) and tested by subcutaneous injection. A solution of the isothiouronium derivative of coumarin in 30 per cent aqueous glycerol was given intraperitoneally. Mercaptohydroxycoumarin was dissolved in water with the addition of $NaHCO_3$.

TABLE 3

DERIVATIVES OF COUMARIN AND ITS SULPHUR-CONTAINING ANALOGUES

No. of cpd.	Compound	Reference to method of preparation	Melting point, °C	
			Found	In literature
1	Coumarin	21	69·5	70
2	4-Hydroxycoumarin	5	206	206
3	7-Hydroxycoumarin	9, 11	224	223–224
4	8-Hydroxycoumarin	19	278	280–285
5	4-Hydroxy-3-bromocoumarin	12	192	192–194
6	4-Hydroxy-3-acetylcoumarin	5	132	132–134
7	4-Hydroxy-3-carboethoxycoumarin	5	101	101
8	7-Hydroxy-4-methylcoumarin	3	185	185
9	6,7-Dihydroxy-4-methylcoumarin	20	271·2	272–274
10	3-Carboethoxycoumarin	16	92	94
11	6-Nitrocoumarin	18	184	185
12	6-Aminocoumarin–HCl	18	169	168–170
13	3-Mercapto-4-hydroxycoumarin	12	208	210
14	3-Isothiouronium-4-hydroxy-coumarin–HBr	12	232	231–234
15	Thiocoumarin	29	100	101
16	Dithiocoumarin*	29	—	—

* Dithiocoumarin was contaminated with a small amount of thiocoumarin.

The results of the tests for protective action are given in Table 4.

In this series of experiments the rats were exposed to X-rays on a 12-tube apparatus at a dose-rate of 130 r/min under conditions in which the dose of 600 r was below the minimum 100 per cent lethal dose for the strain of animals used. The use of oil and glycerol as solvents required supplementary controls with solvents alone. As seen from Table 4, these solvents (under the given conditions of exposure) themselves increased the percentage survival of the rats. From a comparison of the doses it follows that coumarin is more toxic than its

TABLE 4

RESULTS OF A STUDY OF THE PROTECTIVE ACTION OF COUMARIN AND ITS SULPHUR-CONTAINING DERIVATIVES IN RATS

Compound	Dose of substance, mg/100 g	Solvents	Method of administration	Length of time from administration of substance to exposure, hours	Survival rate of rats, numerator—surviving animals, denominator—number taken in experiment	Change in weight of surviving animals by 30 days, % of original	Average duration of life of rats that died, days
Control I	—	—	—	—	3/45	—	12
Control II	—	Oil		2	4/16	120	13
Control II	—	Oil		20	12/38	107	13
Coumarin	10	Oil		20	0/16	—	13
Thiocoumarin	30	Oil	Subcutaneous	2	2/16	122	9
Thiocoumarin	30	Oil		20	15/50	104	13
Dithiocoumarin	25	Oil		2	4/16	102	10
Dithiocoumarin	25	Oil		20	4/16	90	9
Control III	—	30% aqueous solution of glycerol		20	3/10	113	11
3-Isothiouronium-4-hydroxycoumarin (HBr salt)	30	30% aqueous solution of glycerol	Intra-peritoneal	1	2/8	112	10
3-Isothiouronium-4-hydroxycoumarin (HBr salt)	30	30% aqueous solution of glycerol		20	6/20	112	12
3-Mercapto-4-hydroxycoumarin	15	Water with NaHCO₃		0·5	0/16	—	6

Note—Solutions were injected on the basis of 1–1·2 ml./100 g weight. Control I — uninjected; control II — one solvent injected (2 ml. per rat); control III — 30% glycerol injected.

sulphur-containing analogues (thiocoumarin and dithiocoumarin), and when used in a sub-lethal dose it increased the severity of the radiation injury. As regards thio- and dithiocoumarins, and also the isothiouronium derivative, the survival of the treated animals hardly differed from that in the control experiments. Mercaptohydroxy-coumarin proved to be more toxic than thiocoumarin and in the dose used it increased the severity of the radiation injury, not only showing no protective action but actually shortening the life of the rats.

Thus, none of the substances of the coumarin group had protective properties, not even the compound having a free sulphydryl group.

Evidently, for substances to have a protective effect, possession of antioxidant properties and capacity for undergoing oxidation are not the only requirements. Also important are the precise chemical structure, the presence of certain functional groups and their mutual disposition within the molecule.

Thus, certain organic nitrogen thiocompounds, sodium salts of polythionic acids, derivatives of coumarin and its sulphur-containing analogues were synthesized and tested on animals in the study of the radioprotective action of chemical compounds.

Pronounced protective action was noted for 2-amidinothiourea hydrochloride and sodium azidodithioformate dihydrate. A slight protective action was detected in sodium di-, tetra- and pentathionates. No protective action was found in any of the substances of the coumarin series and their sulphur-containing analogues, including compounds having a free thiol group.

Further work on the study of different series of sulphur-containing and other chemical compounds is necessary for solving the problem of the protection of the body from the action of radiation.

REFERENCES

1. KLYUCHNIKOV, N. G., *Handbook of Inorganic Synthesis* (Rukovodstvo po neorganicheskomu sintezu), p. 234, Goskhimizdat (1953).
2. KULBERG, L. M., *Syntheses of Organic Reagents* (Sintezi organicheskikh reaktivov), p. 87, Goskhimizdat (1947).
3. *Syntheses of Organic Preparations* (Sintezi organicheskikh preparatov), Symposium **3**, 218, I. L. Moscow (1952).
4. ABBEG's *Handb. d. Anorg. Chem. Leipzig*, **41**, 554–555 (1927).
5. ANSCHUTZ, R., *Ber.*, **36**, 464 (1903).
6. BACQ, Z. M. and ALEXANDER, P., *Fundamentals of Radiobiology*, Butterworths, London (1955).

7. BAMBERGER, E., *Ber.*, **16**, 1459 (1883).
8. BEILSTEIN's *Handb. d. Org. Chem.*, **2**, 565 (1920).
9. BEILSTEIN's *Handb. d. Org. Chem.*, **3**, 194 (1921).
10. BEILSTEIN's *Handb. d. Org. Chem.*, **3**, 219 (1921).
11. BIZZARI, *Gazz. Chim. Ital.*, **15**, 33 (1915).
12. EISENHAUER, H. R. and LINK, K. P., *J. Amer. Chem. Soc.*, **76**, 1647 (1954).
13. FISCHER, E., *Ber.*, **34**, 433 (1901).
14. FROMAGEOUT, C., *Advances in Enzymology*, **7**, 396 (1947).
15. KHYM, J. X., SHAPIRA, R. and DOHERTY, D. G., *J. Amer. Chem. Soc.*, **79**, 5663 (1957).
16. KNOEVENAGEL, *Chem. Zentr.*, **11**, 1702 (1905).
17. LAWRENCE, J. M. and SMYTHE, C. V., *Arch. Biochim.*, **2**, 225 (1943).
18. MORGAN, MICKLETHWAIT, *J. Chem. Soc. (Lond.)*, **85**, 1233 (1904).
19. PECHMANN, H., *Ber.*, **17**, 932 (1884).
20. PECHMANN, H., *Ber.*, **34**, 423 (1901).
21. PERKIN, *Ber.*, **8**, 1599 (1875).
22. RATHKE, B., *Ber.*, **11**, 962 (1878)
23. ROSENHEIM, A., LEVY, R. and GRÜNBAUM, H., *Ber.*, **42**, 2923 (1909).
24. SHAPIRA, R., DOHERTY, D. G. and BURNETT, W. T., *Radiat. Res.*, **7**, 22 (1957).
25. SMYTHE, C. V., *Arch. Biochim.*, **2**, 259 (1943).
26. SOMMER, F., *Ber.*, **48**, B.11, 1833 (1915).
27. STARKE, M., *J. Prakt. Chem.*, (2), 93, 59 (1916).
28. TAMURA, S., OHKUMA, K. and HAYASHI, T., *J. Agr. Chem. Soc. Japan*, **26**, 410 (1952).
29. TIEMANN, F., *Ber.*, **19**, 1661 (1886).

Effect of β-Mercaptoethylamine on the Formation of Organic Peroxides in the Irradiated Animal

YE. F. ROMANTSEV and Z. I. ZHULANOVA

THE WORK of a number of research workers [8, 9] has shown that during the action of ionizing radiation on the body, compounds are formed which have the properties of organic peroxides. These compounds are formed during exposure and evidently are among the primary products of radiochemical reactions induced by ionizing radiation.

Compounds which are able to protect animals from lethal doses of ionizing radiation act mainly on the primary chemical reactions which develop in the exposed body.

β-Mercaptoethylamine (MEA) is one of the effective prophylactic compounds[5], and naturally considerable interest was aroused in studying the effect of it and other protective substances on the formation of organic peroxides.

The investigation was carried out on white mice. Organic peroxides were extracted by the method of Horgan and Philpot[9]. According to this method, the carcass of the mouse was ground in a blender. The organic peroxides were extracted with butanol and estimated colorimetrically using reduced 2,6-dichlorophenolindophenol.

To construct a standard curve, tetralin hydroperoxide was used. Since in the work of the authors indicated above a number of steps in the procedure are not reported sufficiently, it is expedient to dwell in more detail on the method of preparing the leucocolour and the extraction of the compounds having the character of organic peroxides. (Close adherence to the procedure was required, as small deviations from it led to results which did not agree.)

In all the experiments freshly distilled n-butyl alcohol was used. Anhydrous lithium chloride, used as desiccant, was prepared from commercial chemically pure $LiCl \cdot H_2O$. Distillation of the water was carried out under vacuum on a boiling water bath for 8 hr. The completely dehydrated LiCl was stored in a desiccator over P_2O_5.

To prepare the reduced form of 2,6-dichlorophenolindophenol the dry dye was ground in a mortar with *n*-butanol and a 0·03 M solution prepared. The reducing agent used was ascorbic acid, which was dissolved with heating in *n*-butanol, and a 0·02 M solution was prepared. To 10 ml. of a solution of 2,6-dichlorophenolindophenol in butanol 20 ml. of the solution of ascorbic acid were added.

After the dye had been mixed with the ascorbic acid the solution was heated and filtered through a paper filter. The excess of ascorbic acid was removed by repeated extractions with 5 ml. distilled water. This procedure was repeated until the washing water ceased to reduce the dye. Six extractions were usually sufficient for this.

The reduced 2,6-dichlorophenolindophenol was dried with anhydrous $MgSO_4$ (6 g/30 ml. solution) and stored over $MgSO_4$ in a refrigerator. The dye was filtered before use.

Mice weighing 20–30 g were exposed on an X-ray apparatus of the RUM-3 type in the following conditions: voltage 180 kV, current strength 15 mA, filters 0·5 mm Cu + 1·0 mm Al, dose-rate 62·4–64·5 r/min, total dose 1000 r. The animals were totally exposed on the dorsal surface in Plexiglass cages at a distance of 30 cm.

After exposure the mice were decapitated and placed in a butanol solution of 0·8 M LiCl, 100 ml./30 g mouse. After 10 min the carcass was cut up with scissors under a layer of butanol, and then homogenized in a blender for 20 min (12,000 rev/min), the casing of the blender being cooled in ice. The homogenate obtained was filtered through a paper filter. Two millilitres butanol and 1·0 ml. of a butanol solution of reduced dye were added to 3 ml. of the filtrate and boiled for 10 min on a water bath in a test tube with a ground glass stopper.

The test tube was cooled and after 10 min the colour was measured on a photoelectric colorimeter (FEK–M) with a green filter.

The carcass of an unexposed mouse was treated in the same way. From readings of the extinctions the results of a "blank" experiment in which, instead of the filtrate, a 0·8 M solution of LiCl in butanol was taken, were calculated.

In plotting the standard curve the observed coloration was compared with the coloration given by benzoyl peroxide*. It should be pointed out that to plot the calibration curve, Horgan and Philpot took

---

* The benzoyl peroxide was kindly given to us by D. D. Smolin.

tetralin hydroperoxide, which is more stable than benzoyl peroxide. However, the synthesis of benzoyl peroxide is considerably easier to carry out.

In the first series of experiments the results of Horgan and Philpot (that owing to exposure compounds having the properties of organic peroxides are formed in the body of the mouse) were checked. In this series of experiments the mouse carcasses were ground in the metal beaker of a blender. For each experimental mouse a corresponding

TABLE 1

RESULTS OF A DETERMINATION OF THE LEVEL OF ORGANIC PEROXIDES IN MICE IN A NORMAL CASE AND WITH EXPOSURE (METAL BEAKER)

Date of experiment	Organic peroxides, mg/100 g	
	Normal	After exposure
17/IX 1957	13·5	15·6
19/IX 1957	6·9	13·1
20/IX 1957	9·3	13·1
23/IX 1957	5·7	12·2
24/X 1957	5·7	7·1
28/X 1957	11·1	11·7
20/XI 1957	15·6	17·7
20/XI 1957	11·7	13·6
Average $(\bar{x} \pm \bar{\sigma})$	9·8 ± 1·21	13·0 ± 1·01

$\bar{x}$ — arithmetic mean
$\bar{\sigma}$ — standard deviation

control was set up. Extraction of the organic peroxides and the colorimetric determination were carried out in the course of one day. The tissues were ground at room temperature. The results obtained are given in Table 1.

It follows from the results given in Table 1 that the quantity of compounds having the character of organic peroxides in the body of the mouse increases consistently but slightly as a result of exposure.

As a result of exposure, in the lipid fraction of the mouse carcass on an average 3·2 mg/100 g of peroxide-like compounds are detected.

These, apparently are very labile and liable to breakdown in the presence of traces of metals.

This is indicated by the following observations. If the mouse carcasses are ground in the glass and not the metal beaker of the blender, maintaining all the other conditions of the experiment, the levels of peroxide-like compounds are higher. The results obtained are given in Table 2.

TABLE 2

RESULTS OF A DETERMINATION OF THE LEVEL OF ORGANIC PEROXIDES IN MICE IN A NORMAL CASE AND WITH EXPOSURE (GLASS BEAKER)

Date of experiment	Organic peroxides, mg/100 g	
	Normal	After exposure
8/XII 1957	20·3	29·7
9/XII 1957	17·4	21·7
10/XII 1957	15·1	19·6
12/XII 1957	24·6	25·5
13/XII 1957	21·3	24·6
14/XII 1957	27·2	30·5
16/XII 1957	12·6	13·4
24/XII 1957	18·0	19·3
27/XII 1957	20·7	21·1
10/1 1958	14·6	17·2
15/1 1958	18·2	20·5
Average $(\bar{x} \pm \bar{\sigma})$	19·1 ± 1·22	22·1 ± 1·4

$\bar{x}$ — arithmetic mean
$\bar{\sigma}$ — standard deviation

In comparison with the values obtained when the carcasses were ground in the metallic beaker of the blender, in the latter case (glass beaker) both in the standard and on exposure, the levels of peroxide-like compounds are higher by 1·7–1·9 times. Just as in the first series of experiments, as a result of exposure a slight but constant increase in these compounds of 3·0 mg/100 mg mouse tissue is noted, i.e. almost the same amount as in the first case.

In other papers on the determination of organic peroxides after exposure there are no indications of the effect of the temperature during extraction on the yield of these compounds. According to our observations the temperature factor is of importance. In the experiments given below, the extraction of the lipid fraction was carried out at different temperatures in the glass beaker of a blender. The results obtained are summarized in Table 3.

TABLE 3

RESULTS OF A DETERMINATION OF THE EFFECT OF TEMPERATURE ON THE AMOUNT OF ORGANIC PEROXIDES PRODUCED

No. of experiment	Temperature, °C	Organic peroxides, mg/100 g	
		Normal	After exposure
1	0	8·1	8·6
2	18–20	19·1	22·1
3	65–70	35·6	48·8

Thus, on increasing the temperature during extraction from 0° to 65–70°C, the quantity of compounds having the character of organic peroxides increases 4–6 times.

However, increasing the temperature during extraction to 65–70°C still did not give any certainty that the extraction was complete.

To settle this question, experiments were set up in which multiple extractions of the peroxides were made. The tissues of the control and of the exposed mice were ground at room temperature (18–20°C). The homogenate was then boiled on an oil bath under a reflux condenser for 15 min (118°) and filtered.

An extraction of the residue was repeated 4 times. Each of the 5 extracts was determined colorimetrically, individually.

The results obtained are given in Table 4.

As seen from Table 4, when the carcass homogenate is subjected to prolonged extraction at 118° the amount of peroxide-like compounds both in the control and in the exposed mice increases sharply. If with a single extraction at 18–20° 19·1 mg/100 g was extracted, with five-fold extraction it was 98–111 mg/100 g. From an exposed mouse with a

TABLE 4

RESULTS OF DETERMINATION OF THE AMOUNT OF ORGANIC PEROXIDES IN THE
CARCASS OF A MOUSE BY MULTIPLE EXTRACTION

No. of fraction	Organic peroxides, mg/fraction			
	1st series		2nd series	
	Normal	After exposure	Normal	After exposure
1	8·8	10·1	9·9	11·0
2	8·7	9·7	8·3	8·4
3	4·2	5·3	4·2	4·4
4	1·7	1·7	2·8	2·9
5	1·1	1·2	2·2	2·2
Total	24·5	28·0	27·4	28·9
In 100 g tissue	98·0	107·7	111·0	119
% of normal	100·0	110·0	100·0	107

single extraction at 18–20° 22·1 mg/100 g were extracted, but with multiple extraction 107–119 mg/100 g.

Thus, with multiple extraction it is observed that as a result of exposure the quantity of peroxide-like compounds increased by 9·7–8·0 mg/100 g tissue. If the temperature is increased, however, oxidation may take place during the treatment. Subsequently, therefore, extraction of the lipids was carried out at room temperature.

In the experiments, MEA was injected in a dose of 150 mg/kg, i.e. a dose which, according to the literature[5], exerts considerable prophylactic action. Twenty-five minutes after the injection of MEA the animals were decapitated and the organic peroxides were extracted. These intervals of time were carefully chosen. It should be remembered that in the protection of animals, for example mice, MEA is usually injected 5 min before exposure, but on the other hand the exposure itself in the conditions of the experiment lasted 15 min.

Thus the chosen intervals of time, after which the mice were decapitated, are related to the conditions of exposure of the animals.

The results obtained are given in the diagram.

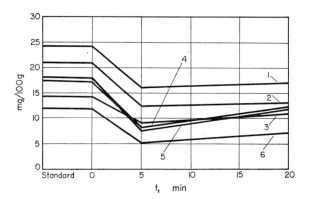

Effect of MEA on the amount of organic peroxides produced: Curves 1–6 correspond to the number of the experiment

It is seen from the diagram that after the injection of MEA in an amount of 150 mg/kg, after 5 min a clearly pronounced reduction in the formation of peroxide-like compounds, about 43·2 per cent compared with the standard, is invariably noted. After 20 min in all cases a steady increase in the amount of organic peroxide-like compounds begins, but their level is still less than the standard on an average by 31·6 per cent.

In a subsequent series of experiments the MEA was injected into the mice, and then after 5 min they were exposed for 15 min to a lethal dose of X-rays. Immediately after exposure the mice were decapitated and a determination of the organic peroxides was carried out.

The results obtained are given in Table 5.

As in the previous series of experiments (see Tables 1 and 2), owing to exposure a constant slight increase in the compounds having the properties of organic peroxides is observed in the lipid fraction.

When MEA is injected for purposes of prophylaxis the amount of organic peroxide-like compounds decreases by 43·2 per cent in comparison with the standard and by 47·2 per cent in comparison with the amount of these peroxides in the exposed mice, i.e. the exposure takes place in conditions which inhibit the formation of these compounds.

TABLE 5

RESULTS OF DETERMINATION OF THE AMOUNT OF ORGANIC PEROXIDES IN MICE
PROTECTED WITH MEA

Date of experiment	Organic peroxides, mg/100 g		
	Normal	After exposure	MEA + exposure
12/XII 1957	24·6	25·5	15·7
20/XII 1957	12·6	13·4	6·6
24/XII 1957	18·0	19·3	10·9
27/XII 1957	20·7	21·1	8·2
10/I 1958	14·6	17·2	7·7
15/I 1958	18·2	20·5	12·7
Average $(\bar{x} \pm \bar{\sigma})$	18·1 ± 1·4	19·5 ± 1·4	10·3 ± 1·2

$\bar{x}$ — arithmetic mean
$\bar{\sigma}$ — standard deviation

The formation of organic peroxides takes place continuously during the normal life of the organism. It is known that a lipoxidase adds on a molecule of oxygen to certain double bonds of unsaturated fatty acids, forming a cyclic organic peroxide, as was first described by Bakh[1].

$$\begin{array}{ccc} \text{R} & & \text{R} \\ | & & | \\ \text{H—C} & & \text{H—C—O} \\ \| & + \text{O}_2 \quad \rightarrow & | \quad | \\ \text{H—C} & & \text{H—C—O} \\ | & & | \\ \text{R} & & \text{R} \end{array}$$

There are also results which show that the enzymatic oxidation of unsaturated fatty acids with conjugated double bonds develops as a chain reaction. As a result of this, peroxide radicals are formed[6].

Reactions of the formation of organic peroxides in the animal body have been insufficiently studied. It is quite probable that they take place to a considerably greater extent than is at present realized. According to the results of investigations on the lipid fraction extracted from a mouse carcass with n-butanol there is a considerable quantity of material which gives reactions characteristic of organic peroxides.

The fact that grinding the mouse carcass in a glass beaker gives twice as much organic peroxide compound as grinding the carcass in a stainless steel beaker deserves attention.

Since the experimental conditions in both cases are the same, it is quite probable that a certain proportion of the organic peroxides decomposes on prolonged contact with metal. It is possible that at the same time a process is being observed which is similar to the catalytic decomposition of hydrogen peroxide by iron ions. The increase in the amount of peroxide-like compounds on heating (see Table 4) may be the result of an increase in solubility of the lipids and of the peroxides which enter into them. In addition, it is quite possible that when the temperature is raised the processes of autoxidation of the labile precursors of the peroxides are intensified.

The fact established by certain investigators[8, 9] that there is an increase in organic peroxides in the tissues of exposed mice was confirmed in our experiments. The character of the solvent is of substantial importance for the separation of these compounds. The most suitable solvent was n-butanol. Butyl acetate, tetrachloroethylene and xylene gave lower results, and absolute ethyl alcohol, according to the results of an experiment, was quite unsuitable.

After exposure of the mice to X-rays Horgan and Philpot observed an increase in the amount of peroxide-like compounds compared with the normal. In our experiments, after a five-fold extraction of the carcasses of exposed mice with n-butanol at 118°C the amount of these compounds increased on an average by only 7–10 per cent, whereas in the experiments of the authors mentioned the peroxide-like compounds increased 2–3 or more times.

It is possible to explain the observed discrepancy:

1. By the difference in time of grinding of the mouse carcasses. Horgan and Philpot ground the tissue for 3 min; in our experiments the tissues were disintegrated for 20 min at 12,000 rev/min. It may be assumed that part of the labile organic peroxide-like compounds decomposes during this time. The period of grinding could not be shortened.

2. By the difference in the organic peroxides used for plotting the calibration curve. Horgan and Philpot[9] used tetralin hydroperoxide; in this case freshly prepared benzoyl peroxide was taken. Yet, in spite

of the quantitative difference, the general regularities in the formation of organic peroxides remained clearly marked.

The formation of organic peroxides as a result of the exposure of lipids was demonstrated in a series of experiments *in vitro*. It may be assumed that as a result of the action of ionizing radiation radicals are formed (for example, of fatty acids) which react with molecular oxygen according to the following scheme[7]:

$$RH \xrightarrow{O\rightsquigarrow} R\cdot + H\cdot$$
$$\text{(fatty acid)}$$

$$R\cdot + O_2 \longrightarrow ROO\cdot$$

$$ROO\cdot + RH \longrightarrow ROOH + R\cdot$$
$$\text{(organic peroxide)}$$

In K. I. Zhuravlev's[2] investigations it was shown that in irradiated fat separated from ox liver, organic peroxides of at least three types are formed.

It is quite possible that a similar process is observed in the exposure of animals to X-rays, but there are not yet any direct indications of this.

The theory put forward by Horgan and Philpot, that the precursor of the organic peroxides in the tissues of exposed mice may be a squalene hydrocarbon with six double bonds, requires experimental checking.

It should be remembered that as a result of exposure, compounds are continuously formed having the properties of organic peroxides, in an amount of 3–8 mg/100 g tissue, or 30–80 mkg/g. In spite of the relatively small quantities, they are probably new compounds and their study must be of special interest.

K. I. Zhuravlev showed in experiments *in vitro* that a number of protective substances prevent the formation of organic peroxides[2].

In our experiments it was established that MEA was capable of causing a reduction in the amount of organic peroxides formed in healthy animals during exposure. On injecting this preparation into mice the exposure of the animals took place in conditions in which there was reduced formation of organic peroxide-like compounds. The mechanism of this action of MEA has not been fully interpreted. It is very probable that, as Ye. F. Romantsev showed, it is connected with the capacity of a number of protective compounds, including MEA, for reducing the consumption of oxygen by animals.

In conclusion it should be stated that the problem of the formation of organic peroxides in the exposed organism has only recently begun to be studied intensively. This is not by chance. A theory exists that these compounds are the origin of a chain reaction[4]. It has recently been shown that in a series of different bacteria, arranged in order of increasing sensitivity to organic peroxides, sensitivity to irradiation also increased along the series. The authors think that the chain reactions begin not only from the formation of lipoid organic peroxides, but develop at the same time in other biological components.

It should be remembered that when the concentration of oxygen in the medium is reduced in experiments *in vitro*, the formation of organic peroxides is reduced[5].

REFERENCES

1. BAKH, A. N., *Zh. Russk. fiz.-khim. obshch.*, **44**, 1–2 (1912).
2. ZHURAVLEV, K. I., Theses of Reports of the Scientific Conference, Pathogenesis, Therapy and Prophylaxis of Radiation Disease (Tezisy dokladov nauchnoi konferentsii, Patogenez terapiya i profilaktika luchevoi bolezni), p. 33, Leningrad (1957).
3. HORGAN, V. and PHILPOT, J., *Radiobiology Symposium* (Bacq and Alexander, eds.), Butterworths, London (1954).
4. TARUSOV, B. N., *Initial Processes of Radiation Disease* (Pervichnye protsessy luchevogo porazheniya), Moscow (1957).
5. BACQ, Z. M. and ALEXANDER, P., *Fundamentals of Radiobiology*, Butterworths, London (1955).
6. BERGSTROM, S. and HOLMAN, R., *Advances in Enzymology*, **8**, 425 (1948).
7. CHEVALLIER, A. and BURG, C., *Radiobiology Symposium* (Bacq and Alexander, eds.), Butterworths, London (1954).
8. DUBOULOZ, P. and DUMAS, J., *Compt. Rend. Soc. Biol.*, 148, 7–8, 705 (1954).
9. HORGAN, V. and PHILPOT, J., Radiobiology Symposium, *Brit. J. Radiol.*, **27**, 313, 63–72 (1954).
10. LATARJET, R., *Ciba Foundation symposium on ionizing radiations and cell metabolism*, 275–296, London (1956).

Possibility of Using Chemical Compounds as Traps for Energy in Protection against Penetrating Radiations

G. Ye. Fradkin

A successful search for protective agents is impossible without a knowledge of the initial processes responsible for the development of the injurious effects of radiation on living material.

Radiation sickness of the living organism is the result of a complicated combination of physico-chemical and biological reactions.

The task of radiobiological research workers consists in differentiating between the more important processes and the accompanying less important ones.

From an analysis of the literature data it is possible to come to the conclusion that at the basis of radiation sickness is a disturbance of the processes of self-reproduction of biological structures, which appears in the first instance as a disturbance in the metabolism of nucleic acids and proteins.

The course and direction of the processes of reproduction of protein structures according to a definite plan characteristic of each type of organism are determined by nuclear and cytoplasmatic nucleic acids, the metabolism and reproduction of which are in turn connected with the metabolism of the proteins. The nucleic acids–proteins system is interdependent and a change of one of the components of this system leads to a breakdown in the processes of self-reproduction.

It is known that during the action of ionizing radiation on the body, the nucleic acids suffer preferential radiation damage compared with the proteins. This situation is confirmed by the fact that the antigenic properties of the protein envelope of bacterial viruses do not change after they have been inactivated by radiation (results of the investigations of D. M. Gol'dfarb, B. N. Il'yashenko and G. Ye Fradkin).

The question naturally arises as to what causes the greater radiosensitivity of nucleic acids. It may be suggested that this depends on the physico-chemical peculiarities of the nucleic acids, the molecules of which absorb the energy of electromagnetic radiations more readily than proteins and other biological compounds. It is known that the molecular extinction coefficient of nucleic acids in the ultraviolet region of the spectrum exceeds the corresponding coefficient for proteins by 60 times.

When γ-rays act on the animal body in lethal doses of the order of 500–600 r, one ten-millionth part of the molecules contained in the tissues is affected by direct ionization. At the same time the yield of free oxidizing radicals, as the calculations of A. V. Savich[4] and other authors show, is not large and amounts to about 0·6 mg OH and about 1·5 mg HO_2 radicals for an animal weighing 20 kg.

The calculated data given above make it difficult to account for the biological action of penetrating radiation merely on the basis of ionization and the resulting formation of free radicals.

The oxygen effect [1, 2, 8, 10] (the existence of which is indisputable) was one of the main arguments in proving the correctness of the idea that the leading role in processes of radiation injury is occupied by oxidizing radicals of the type of HO_2. The existence of this effect results in a decrease in the severity of radiation disease when the level of oxygen is reduced at the time of exposure of the animals and is explained in connection with the mechanism of the initial processes as the result of a reduction in the yield of oxidizing radicals.

However, up to the present it has not been taken into consideration that the oxygen effect in the exposed animal may arise in different ways. It is known, in particular, that the course and the yield of the products of photochemical reactions change in relation to the presence or absence of oxygen in the reaction medium. The presence of oxygen can lead to the formation of stable, irreversible reaction products — photo-oxides (A. N. Terenin[5]). Therefore the mechanism of the protective action of reduced partial pressure of oxygen cannot only be through a reduction in the yield of oxidizing radicals.

It may be thought that when ionizing radiation acts on the body in lethal doses of the order of 500–800 r, injurious chemical reactions are initiated not only by free radicals but also by secondary radiations which arise as a result of the scattering (dissipation) of energy of the primary quanta or particles.

Owing to their physico-chemical peculiarities, nucleic acids absorb the energy of secondary radiations to a greater extent than other biological compounds. As a result of the process of absorption of energy, macromolecules of nucleic acids are activated and drawn into photochemical reactions taking place with the participation of free radicals, which leads to profound and irreversible damage to these structures and to disturbance of the functional activity connected with them. This situation is confirmed by certain literature data concerning the

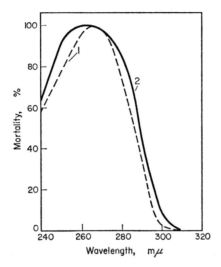

1 — average curve for different bacteria; 2 — curve for golden staphylococcus

FIG. 1. Relationship between the spectrum of the bactericidal action of ultraviolet radiation and the absorption spectrum of nucleic acids

mechanism of radiation injury to the simplest organisms. It has been shown by numerous investigators* that the most lethal part of the spectrum of electromagnetic radiations to microbe cells coincides with the absorption maximum of nucleic acids.

In Fig. 1, results are given showing the relationship between the absorption spectrum of nucleic acids and the spectrum of the lethal action of ultraviolet radiation.

---

* An ample summary of the literature on this question is given in a number of handbooks and monographs: A. MEIER and E. ZEITTS, *Ul'trafioletovoye izlucheniye*, IL, Moscow, 1952; C. ELLIS, A. WELLS, *The chemical action of ultraviolet rays*, New York, 1941; YA. E. NEUSTADT, *Bakteritsidnoye ul'trafioletovoye izlucheniye*, Medgiz, 1955.

It follows from the similarity of the spectra of action and of absorption of energy of electromagnetic radiations that the lethality to cells is connected with the occurrence of harmful photochemical reactions in the nucleic acids. The investigations of Moore and Thomson[9] indicate the high photochemical sensitivity of pyrimidine bases, which are structural elements of nucleic acids. The authors showed that under the influence of ultraviolet quanta uracil, in an aqueous medium, is converted into hydroxyuracil. This reaction gives a relatively high quantum yield (of the order of 0·01). It is possible that photochemical reactions of this type together with other reactions provide conditions for the biological inactivation of nucleic acids during the exposure of the animal body to ionizing radiation. The energy required for the development of a photochemical reaction may be supplied to molecules of nucleic acids also by a non-emissive method of processes of migration of the electronic energy through high molecular-weight compounds[6]. In the living body conditions exist for emissive and non-emissive transfers of energy. Emissive transfers of energy are possible owing to the presence in living systems of molecules with a rigidly constructed skeleton, which are inclined to fluorescence (porphyrins, compounds of the steroid series, tocopherols, flavins and so on).

On the other hand, the presence in living systems of large molecules (proteins, etc.), interconnected into ordered structures by hydrogen bonds, makes conditions favourable for the migration of electronic energy to the site of its absorption (nucleic acids). Without considering these energy transfer processes in detail, it should be pointed out that in both cases the development of chemical reactions injurious to nucleic acids far from the tracks of the primary ionizing quanta or particles is possible.

Chemical traps for energy may decrease the extent of the photochemical reactions which damage the nucleic acids.

Accordingly, there was every reason for setting up experiments on the protection of biological objects from the lethal action of ionizing radiation by means of chemical compounds — energy traps.

In order to simplify the conditions of the experiments and to facilitate analysis of the protective effect, the first series of experiments was carried out on bacterial viruses (bacteriophages), and the second series on laboratory animals. Bacteriophages in the free state (outside the body of the host) do not metabolize and are organized particles consisting of a protein envelope with deoxyribonucleic acid (DNA) enclosed in it and constituting 50 per cent of the total weight of the phage particle. Reproduction of the phage is possible when there is intact DNA which, on penetrating inside the bacterial cell, reorganizes its metabolism in such a way that it reproduces the protein and DNA necessary for the synthesis of the phage particle.

When the DNA is damaged by ionizing radiation the phage loses its capacity for reproduction. Therefore the protection of the phage from the lethal action of radiation may be regarded as a process of prevention of radiation damage to DNA. This circumstance makes the phage a suitable model for studying the protective properties of chemical compounds — energy traps, which have the property of strongly absorbing the energy of electromagnetic radiation in the wavelength region corresponding to the absorption of nucleic acids. Amides of aromatic mercaptoacids and compounds of the pyrimidine series were chosen as energy traps*.

The phage F–1, lysing an intestinal bacillus (strain "600"), was used for the experiments. The phage was exposed to γ-rays from cobalt (^{60}Co) with a dose-rate of about 500 r/min. The exposure was carried out either in Adams' synthetic medium, in which the phage usually reproduces, or in a synthetic medium diluted with aqueous-alcoholic mixtures. An aqueous-alcoholic medium was used when testing the protective action of chemical compounds insoluble in water. It was found by special experiments in a study of the mechanism of the radiation inactivation of bacterial viruses that γ-radiation in doses of 1000, 3000 and 5000 r causes loss of free phage. The maximum inactivation of the phage (90–100 per cent) occurred at different times depending on the size of the exposure dose.

An indication of the inactivation of the phage was the absence or considerable diminution in the extent of its intracellular reproduction.

In the experiments on protecting the phage from radiation injury, solutions (1 ml.) of the chemical compounds were added to a suspension in synthetic medium (4 ml.). The chemical compounds in equimolecular quantities were dissolved in synthetic or in aqueous-alcoholic medium. Corresponding volumes of synthetic or

---

* The amides of the aromatic mercapto acids were synthesized in the laboratory of Academician I. L. Knun'yants, and the pyrimidines in the laboratory of Professor O. Yu. Magidson.

aqueous alcoholic medium were also added to the control samples of phage.

The experimental and control samples of phage were exposed to γ-rays in a dose of 3000 r, after which the tubes with phage were put in a refrigerator for 7–10 days. On the expiration of the indicated period, the concentration of phage particles in the experimental and control samples was determined by the method of agar layers*. The essence of this method consists in that the free phage (1 ml.) comes into contact with the strain of intestinal bacillus which produces it (0·1 ml. washed from a slope inoculated daily) in dilute 0·7 per cent agar cooled to 46°C. The mixture of phage and culture is distributed over the surface of 1·5 per cent agar in a Petri dish. The dishes are put in a thermostat for 4·5 hr. At the end of this period the sterile spots at the base of the growing culture are counted. Each spot corresponds to one phage particle introduced into a bacterial cell.

By comparing the concentrations of phage particles in the experimental and control samples, the degree of protective action of a chemical compound was found. Before setting up the experiments on protection, the toxicity of each chemical compound for phage was studied and those concentrations which did not affect the processes of reproduction were used.

Studies of the protective properties of 3-allyl-4-amino-uracil, the benzylamide of mercaptopropionic acid and the benzylamide of mercaptoisovaleric acid were carried out on suspensions of phage at one concentration.

Experiment 1. A synthetic medium containing phage was subjected to dialysis. After dialysis a suspension of the bacteriophage was poured out into test tubes in portions of 4 ml. To the experimental samples of phage 1 ml. portions of neutral aqueous solutions of 3-allyl-4-aminouracil, the benzylamide of mercaptopropionic acid and the benzylamide of mercaptoisovaleric acid were added. The compounds were used at the same concentration, $0·2 \times 10^{-3}$ moles/l. The experimental and control samples of phage were exposed to γ-rays in a dose of 1000 r. Ten days after exposure the number of surviving phage particles was determined. The results of this experiment are given in Fig. 2.

---

* The scientific co-worker V. F. Ushakova took part in the arrangement of these experiments.

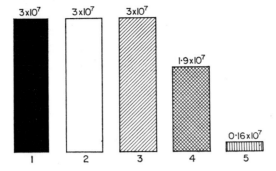

1 — unexposed control; 2 — exposed sample with the addition of 3-allyl-4-aminouracil; 3 — exposed sample with the addition of the benzylamide of mercaptoisovaleric acid; 4 — exposed sample with the addition of the benzylamide of mercaptopropionic acid; 5 — exposed control

FIG. 2. Concentration of phage in 1 ml.

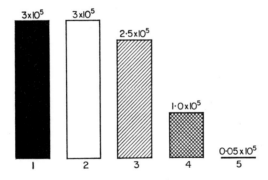

1 — unexposed control; 2 — exposed sample with addition of 3-allyl-4-aminouracil; 3 — exposed sample with the addition of benzylamide of mercaptoisovaleric acid; 4 — exposed sample with addition of benzylamide of mercaptopropionic acid; 5 — exposed control

FIG. 3. Concentration of phage in 1 ml.

Experiment 2. Control and experimental samples in a synthetic medium (with a particle concentration of 3×10^5/ml.) were exposed without preliminary dialysis to γ-rays at a dose of 3000 r. Ten days afterwards the number of surviving phage corpuscles was determined. The results of this experiment are given in Fig. 3.

It follows from the results in Figs. 2 and 3 that 3-allyl-4-aminouracil, when added to the phage before exposure, protects it from inactivation by irradiation. The benzylamide of mercaptoisovaleric acid is close in its effect to allylaminouracil, whereas the benzylamide of mercapto-propionic acid has 2–3 times less protective effect.

The protective effects of benzhydryl ethers of hydroxypyrimidines, on account of their insolubility in water, were studied in aqueous-alcoholic suspensions of phage with a concentration of 10^5 particles per ml.

To the experimental samples of phage, 1 ml. portions of 0.2×10^{-4} molar aqueous-alcoholic solutions of the pyrimidine derivatives were added, and to the control samples of phage 1 ml. portions of aqueous-alcoholic mixture were added.

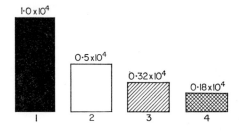

1 — unexposed control; 2 — exposed sample with addition of 2-amino-4(diphenylmethoxy) pyrimidine; 3 — exposed sample with addition of 2,6-diamino-4(diphenylmethoxy) pyrimidine

FIG. 4.   Concentration of phage in 1 ml.

The experimental and control samples of phage were exposed to γ-rays in a dose of 3000 r, and titrated 3 days afterwards. The results of this experiment are given in Fig. 4. The pyrimidine derivatives, added to the bacteriophage before exposure in concentrations 10 times smaller than the concentration of the solutions in experiments 1 and 2, had a pronounced effect, partially protecting the bacteriophage from radiation inactivation.

According to the results of the experiment, 2-amino-4(diphenyl-methoxy)pyrimidine somewhat exceeds the 2,6-diamino compound in its protective action.

In experiments on the chemical protection of phages compounds of cyclic structure were used, with a system of conjugated double bonds capable of strong absorption of the energy of electromagnetic radiations in the ultraviolet region of the spectrum (see Figs. 5–9)*.

It would not, however, be correct to regard intensity of absorption of energy as the only criterion in choosing chemical compounds for the purpose of protection. The protective effect of a chemical compound is

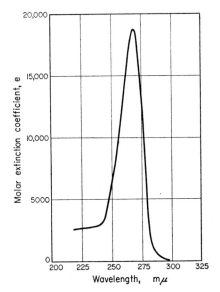

FIG. 5.   Ultraviolet absorption spectrum of 3-allyl-4-aminouracil

determined not only by the absorption factor but also by the fate of the absorbed energy. The energy absorbed by the molecule may be dissipated in three ways: (a) by conversion to the energy of chemical processes; (b) by luminescence; (c) by conversion to heat energy. The last method of conversion of energy is the most desirable for protective compounds.

In testing the protective action of preparations on animals, the same chemical compounds were used as in the experiments with bacteriophages, with the exception of the benzylamide of mercaptoisovaleric acid.

---

* The investigations of absorption spectra were carried out by I. A. Korovina.

E

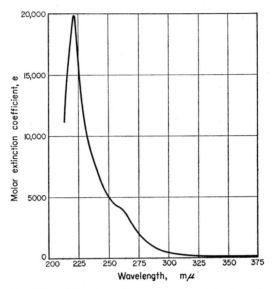

FIG. 6.   Ultraviolet absorption spectrum of the benzylamide of
mercaptoisovaleric acid

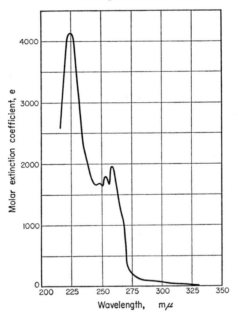

FIG. 7.   Ultraviolet absorption spectrum of the benzylamide of
mercaptopropionic acid

The chemical compounds were previously tested for toxicity to white mice and rats.

Depending on the availability of compounds, the experiments on the protection of animals were done either on white mice (males weighing 20–25 g), or on white rats (males weighing 180–200 g). There were 10 animals in each experimental and control group.

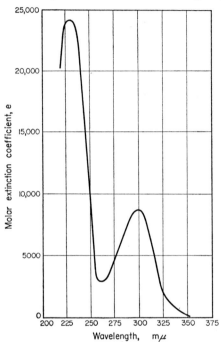

FIG. 8.   Ultraviolet absorption spectrum of
2-amino-4(diphenylmethoxy)pyrimidine

The preparations were injected at different times before exposure, chiefly in the form of suspensions in sunflower oil, previously emulsified with ultrasound at such frequencies of oscillation that no change in the structure of the chemical compounds was possible. The animals were exposed at a cobalt (^{60}Co) γ-source at a dose-rate of approximately 500 r/min.

The effectiveness of the protective action of the preparations was determined from the survival rate of the animals on the 30th day after exposure.

In Tables 1 and 2 the results are given of experiments on white mice, and in Tables 3, 4 and 5, on white rats.

It follows from the data given in Table 1 that the pyrimidine derivatives exert considerable protective effect when they are injected 1 or 2 hr before exposure. β-Mercaptoethylamine, injected 5 min before exposure, gives the same effect.

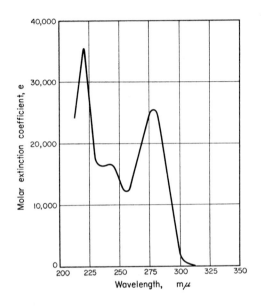

Fig. 9.   Ultraviolet absorption spectrum of
2,6-diamino-4(diphenylmethoxy)pyrimidine

The results of the experiment on testing the protective action of the benzylamide of mercaptopropionic acid when white mice were exposed to γ-rays in a 100 per cent lethal dose (800 r) are given in Table 2.

As seen from Table 2, when animals are exposed to γ-rays in this dosage the benzylamide of mercaptopropionic acid possesses protective action, whereas in these conditions of exposure β-mercaptoethylamine does not give a protective effect.

Table 3 gives results of experiments on the protective action of the benzylamide of mercaptopropionic acid when white rats are exposed to a dose of γ-rays less than 100 per cent lethal (650 r).

From the data given in Table 3 it follows that the benzylamide of mercaptopropionic acid, when injected into rats 1, 3 and 24 hr before exposure, exerts a definite protective action.

Table 4 gives results of experiments on testing the protective action of 3-allyl-4-aminouracil when white rats are exposed to lethal doses of γ-rays.

TABLE 1

RESULTS OF A STUDY OF THE COMPARATIVE PROTECTIVE ACTION OF PYRIMIDINE DERIVATIVES AND β-MERCAPTOETHYLAMINE WHEN GIVEN INTRAPERITONEALLY (EXPOSURE TO γ-RAYS IN A DOSE OF 650 r)

Group (white mice)	Time from injection to exposure, min	2,6-diamino-4-(diphenyl-methoxy)-pyramidine (4 mg per mouse in 0·5 ml. oil)	2-amino-4-(diphenyl-methoxy)-pyrimidine (5 mg per mouse in 0·5 ml. oil)	β-mercapto-ethylamine (3 mg per mouse in 0·5 ml oil)	Control (0·5 ml. oil and 0·5 ml. water)
I	120	4	—	—	—
II	60	6	5	—	3 (oil)
III	45	6	2	—	—
IV	30	5	—	—	—
V	15	3	—	—	—
VI	5	—	—	4	2 (water)

Note—The figures given in the table indicate the survival rate 30 days after exposure.

It follows from Table 4 that 3-allyl-4-aminouracil exerts a certain protective action when rats are exposed to minimum lethal doses (of the order of 700 and 750 r).

In Table 5 the results are given of an experiment on the comparative protective effects of the benzylamide of mercaptopropionic acid and 3-allyl-4-aminouracil when injected intraperitoneally 18 hr before exposure to γ-rays.

It follows from the data given in Table 5 that the benzylamide of mercaptopropionic acid and 3-allyl-4-aminouracil, when injected 18 hr before exposure to γ-rays in a 100 per cent lethal dose, exert a certain protective action corresponding to that of the aminothiols previously known.

On the basis of these experiments it is found that chemical compounds selected on the principle of "traps" for energy protect animals from the action of ionizing radiation under the severe conditions of exposure to a high energy source of γ-rays.

TABLE 2

RESULTS OF A STUDY OF THE PROTECTIVE ACTION OF THE BENZYLAMIDE OF
MERCAPTOPROPIONIC ACID WHEN GIVEN INTRAPERITONEALLY

Group (white mice)	Compound	Dose of compound, mg/mouse	Time from injection of compound to exposure, min	Number of animals surviving	
				After 12 days	After 30 days
I	Benzylamide of mercaptopropionic acid	3	120	7	4
II	Benzylamide of mercaptopropionic acid	3	60	9	4
III	Benzylamide of mercaptopropionic acid	3	30	4	0
IV	β-Mercaptoethylamine	3	5	0	0
V	Sunflower oil (control)	0·5 ml.	60	0	0

Under these conditions, aminothiols of the aliphatic series (β-mercaptoethylamine and so on), when given some minutes before exposure, gave a protective effect coinciding with the effect of chemical compounds "energy traps" which had been given 3, 18 and 24 hr before irradiation.

The mechanism of the protective action of chemical compounds is complicated owing to the variety of processes lying at the basis of radiation sickness.

In the present paper an attempt has been made to throw light on one of the possible sides of this mechanism, connected with the intake of energy of secondary electromagnetic radiations.

The results obtained provide additional material for an understanding of the mechanism of the chemical protection of animals from the lethal action of ionizing radiation.

TABLE 3

RESULTS OF A STUDY OF THE PROTECTIVE ACTION OF THE BENZYLAMIDE OF MERCAPTOPROPIONIC ACID WHEN GIVEN INTRAPERITONEALLY

Group	Compound	Dose of compound, mg/rat	Time from injection of compound to exposure, hr	Number of animals surviving 30 days after exposure
I	Benzylamide of mercapto-propionic acid	20	1	6
II	Sunflower oil (control for Group I)	1 ml.	1	3
III	Benzylamide of mercapto-propionic acid	25	3	10
IV	Sunflower oil (control for Group III)	1 ml.	3	5
V	Benzylamide of mercapto-propionic acid	20	24	3
VI	Aminothiol	55	24	1
VII	Sunflower oil (control for Groups V and VI)	1 ml.	24	0

Note—The animals of groups I, II, III and IV were exposed to γ-rays, and of groups V, VI and VII to X-rays.

Because of the difficulty, in experiments in vivo, of obtaining direct proof of the role of the energy of secondary quanta in the development of radiation disease, in our investigations the means of showing the importance of these processes consisted of chemical compounds having the property of strongly absorbing the energy of electromagnetic radiations in the ultraviolet region corresponding to the absorption spectrum of nucleic acids.

In the model experiments on the protection of bacterial viruses from the inactivating action of ionizing radiation, the relationship of the protective effect to the intensity of absorption by these compounds in the region of the spectrum indicated above was shown for the compounds tested. For example, 3-allyl-4-aminouracil which, in comparison with

TABLE 4

RESULTS OF A STUDY OF THE PROTECTIVE ACTION OF 3-ALLYL-4-AMINOURACIL

Group	Compound	Dose of compound, mg/rat	Time from injection of compound to exposure, hr	Dose of radiation, r	Number of animals surviving	
					After 15 days	After 30 days
I	Allylaminouracil	50	1	700	7	5
II	Allylaminouracil	50	3	700	9	5
III	Sunflower oil (control for Groups I and II)	1 ml.	3	700	4	2
IV	Allylaminouracil	80	3	800	1	1
V	Aminothiol	50	3	800	0	0
VI	Sunflower oil (control for Groups IV and V)	1 ml.	3	800	0	0
VII	Allylaminouracil	150	24	750	6	2
VIII	Starch (control for Group VII)	0·5 ml.	24	750	1	0

Note—Groups I–VI were injected intraperitoneally, and Groups VII and VIII orally.

the benzylamide of mercaptopropionic acid absorbs more strongly in this region of the spectrum gives total protection of the bacteriophage, whereas the second compound protects only partially. In experiments on the protection of animals using the same compounds such a relationship could not be demonstrated. It should be pointed out that in experiments on animals the protective effect is a function of a whole series of biological factors: of the rate of distribution and penetration of the chemical compounds into the cells, of the rate of removal of these compounds from the body, of their inclusion in the metabolism, and of their accompanying pharmocological effects, and so on. Therefore in those cases where the best conditions for the manifestation of the protective effect have not yet been found, a number of biological factors may be overriding the physico-chemical action of the protective compounds.   Whatever   the   circumstances,   these   experiments

TABLE 5

RESULTS OF A STUDY OF THE COMPARATIVE PROTECTIVE ACTION OF THE BENZYLAMIDE OF MERCAPTOPROPIONIC ACID, 3-ALLYL-4-AMINOURACIL AND AN AMINOTHIOL

Group (white rats)	Compound	Dose of compound, mg/rat	Time from injection of compound to exposure, hr	Number of animals surviving 30 days after exposure
I	Benzylamide of mercapto-propionic acid	25	18	3
II	Allylaminouracil	50	18	3
III	Aminothiol	50	18	3
IV	Sunflower oil (control for Groups I, II and III)	1 ml.	18	0

demonstrate indisputably the existence of protection of animals (white mice and rats) from the lethal action of γ-rays of high energy by means of chemical compounds selected on the principle of energy "traps".

It is also found that, besides thiol compounds, chemical compounds of another class, not having easily oxidizable sulphydryl groups, exert protective action.

In conclusion, one should consider the question of whether the mechanism of the protective action of aminothiols of the aliphatic series could be explained from the standpoint of intake of energy of secondary radiations. According to V. Noies and V. Bekel'khaid[3], the photo-oxidation of mercapto-compounds proceeds extensively during exposure to short-wave ultraviolet radiation (with wavelength less than 280 mμ).

Consequently, aminothiol compounds, in using part of the secondary energy on the oxidation of their SH-groups, may decrease the extent of the photochemical reactions which damage the nucleic acids.

CONCLUSION

Chemical compounds (mainly derivatives of pyrimidine bases) which have a clearly pronounced property of absorbing the energy of electromagnetic radiations in the ultraviolet region of the spectrum, corresponding to the absorption of nucleic acids, give a small protective effect when injected into animals before they are exposed to γ-rays in lethal doses.

For chemical compounds of the pyrimidine series to show a protective effect, the presence of sulphydryl groups in the molecule is not obligatory.

REFERENCES

1. GRAYEVSKII, E., *Usp. sovremennoi biologii*, **37**, 2 (1945).
2. GRAYEVSKII, E. and OGINSKAYA, G., *Dokl. Akad. Nauk SSSR*, **89**, 737 (1953).
3. NOIES, V. and BEKEL'KHAID, V., *Methods for the Photochemical Synthesis of Organic Substances* (Metody fotokhimicheskogo sinteza organicheskikh veshchestv), Izd. I. L. Moscow (1951).
4. ROMANTSEV, YE. F. and SAVICH, A. V., *Chemical Protection from the Action of Ionizing Radiation* (Khimicheskaya zashchita ot deistviya ioniziruyushchei radiatsii), Part I, Medgiz (1958).
5. TERENIN, A. N., *Photochemistry of Dyes and Related Organic Compounds* (Fotokhimiya krasitelei i rodstvennykh organicheskikh soyedinenii), Akad. Nauk SSSR, Moscow, Leningrad (1947).
6. ALEXANDER, P. and CHARLESBY, A., *Nature*, **173**, 508 (1954).
7. GRAY, L. N., *Acta Radiol.*, **41**, 1 (1954).
8. LEA, D., *Actions of Radiations on Living Cells*, Cambridge (1955).
9. MOORE, A. M. and THOMSON, C. H., Photocomposition of Pyrimidine Compounds, *Progress of Radiobiology*, p. 75, London (1956).
10. READ, I., *J. Radiol.*, **25**, 300 (1952).

PART II

REMOVAL OF
RADIOACTIVE ISOTOPES
FROM THE BODY

General Information

V. S. Balabukha

The search for methods of removing radioactive isotopes from the body is no less important a problem than the chemical prophylaxis of radiation sickness. The possibility of the radioisotopes entering by the oral route or through the respiratory organs on direct contact with active substances is far from excluded. This problem has become more real in recent years in connection with the increased contamination of the atmosphere, soil and plants with radioactive fall-out resulting from the repeated testing of atomic weapons. Those radioactive isotopes which are localized mainly in the bone tissue and are retained for a long period present a special danger. It is therefore natural that the attention of research workers should be attracted to these problems.

An extensive literature has already been collected which throws light mainly on questions of the distribution of radioactive isotopes in the body, the time they are retained in the organs and tissues, and the rate and means of excretion.

There are also studies dealing with the possibility of eliminating these isotopes by means of chemical substances which form complex compounds with them or adsorb them fairly firmly (see, for example, the reviews by Schubert and Balabukha and Fradkin*).

However, very little attention has been paid to the state of radio-isotopes in the blood and in bone tissue, or to the factors involved in their deposition at certain preferred sites. There are almost no systematic comparative physico-chemical and biological studies of the stability of the complexes formed from radioactive isotopes and complex-forming agents, *in vitro* and under the conditions obtaining *in vivo*. The latter are necessary from a practical point of view for the selection of the most effective complexing agents.

The results of experimental investigations devoted to solving these problems are below.

---

* J. Schubert, *Ann. Rev. Nucl. Sci.*, **5**, 369 (1955); V. S. Balabukha and G. Ye. Fradkin, *Accumulation of radioactive elements in the body and their removal* (Nakopleniye radioaktivnykh elementov v organizme ikh vyvedeniye), Medgiz, Moscow (1958).

Physico-Chemical (Chromatographic) Study of the Effectiveness of Certain Complexing Agents

L. I. Tikhonova and L. M. Razbitnaya

In the problem of the therapy of radiation disease caused by radioactive isotopes in the body, questions connected with the elimination of these isotopes with the excreta are of paramount importance. At the same time the possibility of the formation of complex compounds of radioisotopes with different organic complexing agents plays an important part.

The complexing agents must be non-toxic and must form water-soluble complex compounds with radioactive isotopes, these compounds being readily diffusible and stable under the conditions existing within the body. The object of this work was to search for the most effective complexing agents for removing radioactive elements from the body, on the basis of the chromatographic method worked out earlier[1].

The investigations were carried out with the following radioactive isotopes: [89, 90]Sr, [91]Y, ([175]Yb), [210]Po, [95]Zr.

The choice of these elements was not by chance, since their production and ever-increasing use both in industry and in numerous investigations gives rise to the danger of their entering the human body.

In relation to their position in D. I. Mendeleyev's periodic system, the elements studied have different capacities for complex-formation, and therefore elimination from the body under the influence of various complexing agents will have its own peculiarities in each case. It should particularly by borne in mind that of the elements indicated above the complex-forming properties are least pronounced in Sr[2]. To study the composition and stability of the complex compounds, different chemical and physico-chemical methods are used. This investigation is based on the simplest chromatographic method, which is fairly accurate for the determination of the stability of the complexes formed, and consists in the conversion of the cationic form of the element to the anionic on a column of cationite*:

$$Me^{n+} + A^{m-} \rightarrow MeA^{(m-n)-}, \text{ where } m-n>0$$

This conversion is associated with the formation by the element of a complex with the organic complexing agent. If the complex anion formed is stable, it quickly leaves the cationite column, i.e. the rate of washing out of the chromatographic zone (V of zone) is almost equal to the rate of movement of the washing solution (V of solution). When an unstable complex anion is formed the element being studied does not appear in the filtrate for a long time, i.e. V of zone $\ll V$ of solution.

Thus, V of the zone depends on the stability of the complex compound and on its instability constant K_i. The chromatographic method for the comparative determination of the complex-forming capacities of organic compounds has previously been described in detail[1].

In practice the method amounted to the determination of v_{max} of the elution curve, which gives the degree of stability of the complex compound. The more stable the complex compound is, the smaller is the value of v_{max}, because the latter is inversely proportional to V of the zone. The elution curve of an element from a column of cationite is given below (see figure).

By this method the complex-forming capacity of a number of organic compounds was determined in relation to the elements indicated above (the concentrations of the solutions of complexing agents were 0·003, 0·05, 0·5 mole/l., pH 6·7). The strongly acidic cationite KU-2 was used in the experiments.

The results are given in Tables 1–4.

These show that for Y cyclohexanediaminetetraacetic and cyclopentanediaminetetraacetic acids may be recommended, since with this element they form more stable complex compounds than ethylenediaminetetraacetic and phenylethylenediaminetetraacetic acids. The stability of the complex with phenylethylenediaminetetraacetic acid is similar to that of the complex with ethylenediaminetetraacetic acid. Complexes of cysteinyl-β-hydroxypropionic and cysteinyllactic acids with [91]Y are less stable than complexes with EDTA.

---

* [cation-exchange resin — Ed.]

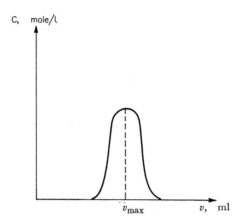

Curve of elution of a radioactive element from a column of cationite
by a complexing agent

TABLE 1

v_{max} (IN ML.) OF THE ELUTION CURVE OF Sr^{+2} FROM A COLUMN OF
CATIONITE KU–2

Complexing agent (acid)	Concentration of solution of complexing agent, mole/l.		
	0·003	0·05	0·5
Cyclohexanediaminetetraacetic	250	—	—
Cyclopentanediaminetetraacetic	270	—	—
Ethylenediaminetetraacetic	320	—	—
Uramildiacetic	1075	—	—
Citric	—	500	—
Tartaric	—	720	—
Sulphosalicylic	—	—	280

TABLE 2

v_{max} (IN ML.) OF THE ELUTION CURVE OF Y^{3+} (Yb^{3+}) FROM A COLUMN
OF CATIONITE KU–2

Complexing agent (acid)	Concentration of solution of complexing agent, mole/l.	
	0·003	0·006
Cyclohexanediaminetetraacetic	0·50	—
Cyclopentanediaminetetraacetic	0·50	—
Ethylenediaminetetraacetic	0·7	—
Phenylethylenediaminetetraacetic	0·7	—
Cysteinyl-β-hydroxypropionic	1·5	—
Cysteinyllactic	2·7	—
Uramildiacetic	6·5	—
Anthranilinodiacetic	7·2	—
Ethylenediamine-bis-isopropylphosphinic	—	2·02
Ethylenediamine-bis-benzylphosphinic	—	1·62
Ethylenediamine-bis-propylphosphinic	—	1·53
Ethylenediamine-bis-methylphosphinic	—	1·27
Aminoisopropylphosphinic*	1·1	—
Hydroxybenzylphosphinic*	1·9	—

* Concentration — 0·05 M.

Anthranilinodiacetic and uramildiacetic acids, and derivatives of ethylenediamine-bis-phosphinic and phosphinic acids (ethylenediamine-bis-methylphosphinic, ethylenediamine-bis-propylphosphinic, ethylenediamine-bis-isopropylphosphinic, aminoisopropylphosphinic and so on), are considerably weaker than EDTA in their capacity for forming complexes with Y^{3+}. [3] There have recently appeared in the literature reports of stronger complexing agents which evidently have prospects for use with Y^{3+}. [4, 5]

As seen from Table 3, unithiol, cyclopentane-, ethylene- and cyclohexanediaminetetraacetic acids form stable complex compounds with Po.

TABLE 3

v_{max} (IN ML.) OF THE ELUTION CURVE OF PO FROM A COLUMN OF
CATIONITE KU–2

Complexing agent (acid)	Concentration of solution of complexing agent, 0·003 mole/l.
Unithiol (2,3-dimercaptopropylsulphonic)	$\leqslant 10$
Cyclopentanediaminetetraacetic	$\leqslant 10$
Ethylenediaminetetraacetic	$\leqslant 10$
Cyclohexanediaminetetraacetic	$\leqslant 10$
Cysteinyl-β-hydroxypropionic	16
Cysteinyllactic	22
Ethylenediaminetetraacetic (Ca, Na-salt)	300
Loretin	> 2000

TABLE 4

v_{max} (IN ML.) OF THE ELUTION CURVE OF Zr^{4+} FROM A
COLUMN OF CATIONITE KU–2

Complexing agent (acid)	Concentration of solution of complexing agent, 0·003 mole/l.
Cyclohexanediaminetetraacetic	$\leqslant 5$
Cyclopentanediaminetetraacetic	$\leqslant 5$
Ethylenediaminetetraacetic	$\leqslant 5$
Uramildiacetic	$\leqslant 5$

For Zr^{4+}, an analogue of Pu, the effective complexing agents are cyclopentane-, cyclohexane-, ethylenediaminetetraacetic and uramildiacetic acids.

Sr^{2+} forms relatively stable complexes with cyclohexane-, cyclopentane- and ethylenediaminetetraacetic acids. However, their use in the removal of radioactive strontium from the body is complicated by the fact that Ca^{2+}, which is found in the blood and tissues in considerable quantity, forms no less stable compounds with these same complexing agents and is thus a competitor of the Sr^{2+}. In order to eliminate Sr^{2+} from the body, it is therefore necessary to look for such complexing agents as can form with Sr complex compounds which are more stable than those with Ca^{2+}.

REFERENCES

1. SENYAVIN, M. M. and TIKHONOVA, L. I., Chromatographic method for the comparative determination of the stability of complex compounds, *Zh. neorg. khim.*, **1**, No. 12, 2772 (1956).
2. GRINBERG, A. A., *Introduction to the chemistry of complex compounds* (Vvedeniye v khimiyu kompleksnykh soyedinenii), Goskhimizdat, Moscow, Leningrad (1951).
3. KABACHNIK, M. I., MEDVED', T. YA., KOZLOVA, G. K., BALABUKHA, V. S. and TIKHONOVA, L. I., *Izv. Akad. Nauk*, No. 9, 1070 (1958).
4. CATSCH, A., DU KHUONG, LE, *Nature*, **180**, 4586, 609 (1957).
5. KROLL, H., KORMAN, S., SIEGEL, E., HART, H. E., ROSOFF, B., SPENSER, H. and LASZLO, D., *Nature*, **180**, 4592, 919 (1957).

Characteristics of the State of the Radioactive Isotopes ^{89}Sr, ^{91}Y and ^{144}Ce in the Blood

L. M. RAZBITNAYA and V. S. BALABUKHA

THE FORM in which any radioactive element is found in the blood and the form in which it is eliminated from the body, are of considerable practical interest from the point of view of an analysis of its behaviour in the body and of checking the effectiveness of synthetic preparations intended for accelerating the elimination of these substances. The metabolism of radioactive elements, their accumulation in the organs and tissues, and the nature of the elimination process depend on a whole series of factors, but they are also determined by the chemical and physico-chemical properties of the substances themselves. It must be taken into account that when inorganic compounds get into the body their physico-chemical condition may be changed by the blood, which is a complex buffer system containing proteins, various inorganic ions and anions of organic acids, capable of forming complexes with cations.

Literature data on the forms of binding of radioactive elements in the body are comparatively few and rather contradictory. The investigations are mainly concerned with the distribution and fixation of various radioactive isotopes in the organs in relation to methods of introduction and the quantity introduced, but they throw little light on the question of chemical binding and reaction of the isotopes with the tissues, particularly with the blood. Altogether it is possible to indicate only a few papers in which this question is touched on to any extent. In the literature there are, for example, data on features of the metabolism of radioactive caesium and rubidium in the blood. Thus, Hood and Comar[6] detected a gradual decrease in the concentration of caesium in the plasma on account of its entry into cells, which 20–30 hr after injection contained 15 times more Cs than the plasma. Love and Burch[10] found that ^{134}Cs and ^{86}Rb are concentrated, like ^{42}K, in the red blood corpuscles, and their metabolism in the erythrocytes is almost the same as the metabolism of potassium[2].

Regarding the state of Ca in the plasma, some authors hold the opinion that there is a complex equilibrium system

$$Ca \rightleftarrows proteins \rightleftarrows Ca^{++} \rightleftarrows CaHPO_4 \rightleftarrows colloidal\ phosphate$$
$$\updownarrow$$
$$citrate$$
$$malate$$

Some confirmation was obtained in investigations with the radioactive isotope of calcium.

Armstrong, Johnson, Singer, Lienke and Premer[1], when carrying out experiments *in vitro* with ^{45}Ca, with the object of determining the effect of the binding of Ca to the proteins of the serum on its capacity for passing through the walls of the capillaries, came to the conclusion that this binding does not prevent such movement. They therefore consider that a mobile equilibrium exists between ionic calcium and the calcium bound to plasma protein. Zilversmit and Hood[9], however, in studying the physico-chemical state of certain radioactive isotopes in blood plasma by paper electrophoresis found that ^{45}Ca present in the plasma always migrates in the direction of the cathode, i.e. it is in an ionic state, whereas proteins move towards the anode.

Indications of the nature of the interaction of radioactive strontium with the blood are, in general, absent from the literature, if Kity's data are not counted[8]. On the basis of the stability constants with organic acids he calculated that only 0·3 per cent of the Sr is combined with citric acid in the blood, but that the bulk of it is in the cationic form. Information on the nature of the binding of rare earth and heavy elements in the blood is also very limited. There are some indirect data with regard to yttrium. For example, Kawin[7], when studying the mechanism of the distribution and excretion of radioyttrium by means of paper chromatography, came to the conclusion that there is a two-phase process of excretion, consisting of:

(1) the direct elimination of injected inorganic yttrium;
(2) the elimination of an amino-acid complex of yttrium from the plasma and tissues.

For an investigation of the physico-chemical state of radioactive yttrium in the blood, Zilversmit and Hood used paper electrophoresis of blood plasma containing ^{91}Y. They came to the conclusion that in the conditions of the experiment (pH 8·5) this radioactive element exists in two states, mobile and immobile. Dudley[3] stated that the deposition of radioactive yttrium in the different tissues depends on the complex-forming agent with which the yttrium was combined, and in what proportions, before introduction into the body. In experiments *in vitro* the author also showed that in the complexed form yttrium dialyses through cellophane membranes.

Campbell and Talley[2] studied the distribution of polonium in the blood of rats and dogs and found that 92 per cent of the polonium is combined with the red blood corpuscles and 4 per cent with the white. Thomas and Stannard[12] show that 96 per cent of the polonium is combined with haemoglobin.

The present investigation is devoted to a study of the behaviour of radioactive strontium, yttrium and cerium in the blood.

Strontium, like calcium, does not belong to that group of elements with a strongly pronounced complex-forming capacity. On entering the blood, the radioisotope of strontium does not form sufficiently stable compounds with organic acids, particularly as the concentration of the latter is small (2–15 mg per cent). It is more probable that it circulates in the blood mainly in a non-complexed form or combined very unstably with proteins.

Yttrium and cerium, which belong to the group of rare earth elements, are characterized by a fairly strongly pronounced capacity for complex-formation and may enter into a more stable association with proteins.

It is necessary to take into account the fact that at the pH of the blood yttrium and cerium form hydroxides, i.e. they may exist, at least partially, in the colloidal state, provided one or more normal components of the blood do not prevent this by complex-formation.

In order to study the character of the binding of radioactive isotopes of strontium, yttrium and cerium in the blood, the method of electrophoresis was used. The principle of the method used consists in that if the radioactive element forms a complex compound with a certain organic acid, it is found in solution in the form of a complex anion and under the action of a direct electric current it will move towards the anode. If, however, the radioactive element is not connected with an organic acid, i.e. it exists in the form of a positively charged cation, on electrophoresis it will migrate in the direction of the cathode. By determining the amount of radioactive element in the cathode and anode solutions, it is possible to judge whether the radioactive element is combined with the acids indicated in a complex compound or remains free, i.e. is present in a cationic form. To carry out the electro-migration investigations the apparatus used earlier by Seaborg[11] for studying the behaviour of Np and Pu in solutions of certain acids (see figure) was used.

The apparatus consists of three glass sections connected by fine-pored membranes of Schott glass. It is provided with two platinum electrodes sealed into thin glass tubes. In the middle section the liquid to be studied is placed, and in the side sections, each of which consists of two bends, are the solutions of acids or NaCl of known concentration. To prevent the possible adsorption of radioactive isotopes by the walls and especially by the pores of the membranes the apparatus is previously saturated with a stable isotope of strontium if working with radioactive strontium and with lanthanum if working with yttrium and cerium. In order to prevent convection currents and mixing of the solutions, the electrophoresis vessel is placed in cold water and then connected with the electric circuit (current strength 5–10 mA).

At the end of the experiment the liquid from the anode and cathode vessels is transferred to test tubes with ground glass stoppers. Aliquot

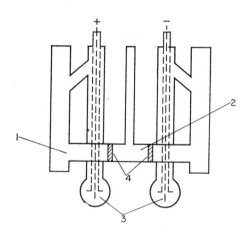

1 — solution of NaCl; 2 — liquid being studied; 3 — Pt-electrodes;
4 — membranes

Apparatus for electrophoresis

portions are carefully evaporated on stainless steel planchets and counted.

The electromigration behaviour of ^{89}Sr and ^{91}Y in solutions of citric, lactic and glutamic acids, taken in concentrations corresponding to their concentration in the blood (citric 2–3 mg per cent, lactic 15 mg per cent and glutamic 3–4 mg per cent), was studied by this method.

A solution of the acid, to which the radioactive isotope of strontium or yttrium with an activity of 5000–6000 counts/min had been previously added, was introduced into the central part of the apparatus.

TABLE 1

RESULTS OF A STUDY OF THE ELECTROMIGRATION OF ^{89}Sr IN SOLUTIONS OF ORGANIC ACIDS

Acid	Concentration of acid, mg %	^{89}Sr content of solution, % of total		
		In cathode	In central section	In anode
Citric	2–3	99·8	None	Traces
Glutamic	3–4	99·5	None	None
Lactic	15	100	None	None

The same acid was introduced into the anode and cathode bends, but without the radioactive isotope. The possibility of the oxidation of citric acid at the anode was previously investigated by the method of A. P. Safronov[13] and it was found that under the conditions of the experiment at a current strength of 5–10 mA no noticeable oxidation takes place.

The results of the investigations of the electromigration behaviour of microquantities of strontium and yttrium in solutions of organic acids when the electrophoresis is continued for 18 hr are given in Tables 1 and 2.

As seen from the tables, the radioactive isotope of strontium in solutions of acids does not enter into stable complex formation. Under these conditions the yttrium isotope appears to be complexed with citric acid to the extent of about 2 per cent. An increase in the concentration of acid gives a higher percentage combination of the radioactive isotopes. This was shown, for example, for ^{89}Sr and ^{91}Y and citric acid of different concentrations. Thus, an increase in the concentration of citric acid to 0·003 M leads to combination with the organic acid into a complex anion of 10 per cent of the ^{91}Y and 6 per cent of the ^{89}Sr. When the concentration of acid is increased to 5 per cent the extent of incorporation of ^{91}Y into a citrate complex amounts to 50 per cent.

TABLE 2

RESULTS OF A STUDY OF THE ELECTROMIGRATION OF ^{91}Y IN SOLUTIONS OF ORGANIC ACIDS

Acid	Concentration of acid, mg %	^{91}Y content of solution, % of total		
		In cathode	In central section	In anode
Citric	2–3	98	None	2
Glutamic	3–4	100	None	None
Lactic	15	100	None	None

Note—The values in Tables 1, 2 and 3 and 5 are the average of the results of 2–3 experiments. The variation does not exceed 5 per cent.

In the second part of the work the combination of the radioactive isotopes indicated with proteins in the blood was studied. The investigations were carried out with serum which in the first series of experiments was obtained from the blood of healthy animals, and in the second from the blood of animals after they had been injected with the radioactive isotopes ^{89}Sr, ^{91}Y or ^{144}Ce at 0·1 mc/kg. In the first case the radioactive isotopes were introduced into the serum *in vitro* in the form of salts of inorganic acids ^{89}Sr(NO$_3$)$_2$ and ^{91}YCl$_3$ at a level of 5–6 thousand counts/min /1 ml. In the injected animals the blood was taken from the aural vein 0·5–1 hr after injection, i.e. at a time when the radioactive isotopes are still in the blood.

For the electrophoresis an apparatus similar to that described earlier was used, except that the cathode and anode membranes of Schott glass were replaced by Cellophane ones with pores 1 mμ in size. These membranes did not let through protein molecules from the serum, which was placed in the middle section. If the radioactive isotope being studied was firmly bound to the protein molecules, on electrophoresis it remained entirely in the central section of the apparatus. If, however, the radioactive isotope was in the form of a complex anion or free cation or its binding with the proteins of the serum was labile, then under the influence of a direct electric current it moved correspondingly into the anode or cathode field.

TABLE 3

ELECTROPHORESIS OF BLOOD SERUM CONTAINING RADIOACTIVE ISOTOPES

Element	Preparation of of serum	Duration of electrophoresis, hr	Radioactive isotope content of solution, % of total		
			In cathode	In central section	In anode
^{89}Sr	From blood of a healthy rabbit	40	99·8	—	0·2
^{89}Sr	From blood of a rabbit after injection of radioactive isotope	40	99·0	—	1·0
^{91}Y	From blood of a rabbit after injection of radioactive isotope	12	Traces	100	Traces
^{91}Y	From blood of a rabbit after injection of radioactive isotope	30–40	99·8	None	Traces
^{91}Y	0·3% NaCl	30–40	99·5	None	Traces

The conditions of the experiment are similar to those described earlier. In the middle section was placed the serum to be investigated, and in the cathode and anode sections a 0·3 per cent solution of NaCl. The results of the experiments are given in Table 3.

As seen from Table 3, the radioactive isotope of strontium passes through the Cellophane membranes and is found almost entirely in the cathode solution. Consequently, it is not combined in significant quantities either with the organic acids which are in the blood serum, or with the proteins.

On electrophoresis of the serum, the yttrium isotope remains in the central part of the apparatus. On continuing the electrophoresis (for 30–40 hr) and with a gradual change of pH to 2, the radioactive isotope of yttrium gradually accumulates in the cathode solution. A similar picture is observed in the blank experiment, i.e. in the electrophoresis of ^{91}Y in a 0·3 per cent solution of NaCl in the absence of protein.

TABLE 4

ELECTROPHORESIS OF BLOOD SERUM CONTAINING ^{91}Y
(Duration of electrophoresis 40 hr)

pH	Radioactive isotope content of solution % of total		
	In cathode	In central section	In anode
2	100	None	Traces
5·5	99	None	Traces

In order to elucidate the nature of the binding of micro-amounts of yttrium in blood serum, special experiments were conducted. Serum, taken from the blood of a rabbit after injection of the radioactive isotope, was subjected to electrophoresis at pH 2, i.e. in circumstances where the possibility of the formation of a hydroxide was excluded. In this case the movement of radioactive yttrium towards the cathode began immediately and continued to the end.

Electrophoresis of serum containing the radioactive isotope of yttrium at pH 5·5 was also carried out.

The results of these experiments are given in Table 4.

From a comparison of the results in Tables 3 and 4, the decisive influence of the pH of the medium on the behaviour of microquantities of ^{91}Y is evident. The tendency of yttrium to hydrolysis must be taken into consideration when discussing the results of the electrophoresis. For example, Grogan and Oppenheimer[4] show that the large percentage of radioactive chromium linked with proteins, and found by Gray and Sterling[5], is explained by the hydrolysis of the non-diffusible chromium.

An attempt was made to separate the proteins of the blood serum which contained the radioactive isotope of yttrium, by paper electrophoresis. The method gives a good separation of the serum proteins into four fractions, but no fraction contains such an amount of radioactive yttrium as to indicate that it is mainly associated with that fraction. The radioactive isotope is more or less uniformly distributed over the whole strip of paper. Electrophoresis on paper in

TABLE 5

ELECTROPHORESIS OF BLOOD SERUM CONTAINING ^{144}Ce
(Duration of electrophoresis 40 hr)

pH	Radioactive isotope content of solution % of total		
	In cathode	In central section	In anode
5·5	95	Traces	4·5
7·4	95·1	Traces	4·9

the case of yttrium is complicated by a number of secondary phenomena: by the reaction of the radioactive element with the paper, by capillary movement of the solution over the strip of paper, and by the high pH of the buffer solution (8·5).

Comparison of the results obtained enables the hypothesis to be put forward that radioactive yttrium is found in the blood serum either in the form of radiocolloids or in the form of pseudocolloids, i.e. adsorbed on the protein molecules.

The fact that even in mild conditions of electrophoresis (current strength 5–10 mA) there is sufficient change of pH of the medium for the gradual removal of yttrium from the central section of the apparatus, containing the proteins of the serum, and its accumulation in the cathode solution, indicates that there is no stable chemical binding with the proteins of the serum.

Similar experiments were carried out with the radioactive isotope of cerium. The results are given in Table 5.

It is seen from the data given that ^{144}Ce migrates mainly to the cathode solution and only 5 per cent is bound into a complex anion. In contrast to ^{91}Y the movement of ^{144}Ce towards the cathode is observed in the first hours of electrophoresis. No stable binding of ^{144}Ce with proteins of the serum has been found.

From these investigations it may be concluded that the radioactive isotope of strontium in the blood serum is not combined with the organic acids in noticeable amounts but is found in the form of a cation or bound very unstably with molecules of protein. No stable chemical binding is observed between the radioactive isotopes of yttrium and cerium and the proteins of the blood serum. They evidently exist in the serum in the form of radiocolloid or pseudocolloid.

REFERENCES

1. ARMSTRONG, W. D., JOHNSON, J. A., SINGER, L., LIENKE, R. J. and PREMER, M. L., *Amer. J. Physiol.*, **171**, 64 (1952).
2. CAMPBELL, J. E. and TALLEY, L. H., *Proc. Soc. Exptl. Biol. Med.*, **87**, 1, 221 (1954).
3. DUDLEY, H. C., *J. Lab. Clin. Med.*, **45**, 5, 792 (1955).
4. GROGAN, CH. H. and OPPENHEIMER, H., *Arch. Biochim. Bipohys.*, **56**, 1, 204 (1955).
5. GRAY, S. J. and STERLING, J., *J. Clin. Invest.*, **29**, 16014 (1950).
6. HOOD, S. L. and COMAR, C. L., *Arch. Biochim. Biophys.*, **45**, 423 (1953).
7. KAWIN, B., *Arch. Biochim. Biophys.*, **45**, 230 (1953).
8. KITY, S. S., *J. Biol. Chem.*, **142**, 181 (1942).
9. ZILVERSMIT, D. B. and HOOD, S. L., *Proc. Soc. Exptl. Biol. Med.*, **84**, 3, 573 (1953).
10. LOVE, W. D. and BURCH, G. E., *J. Lab. Clin. Med.*, **41**, 351 (1953).
11. SEABORG, G., *The Transuranium Elements*, Part 1, p. 358 (1949).
12. THOMAS, R. G. and STANNARD, J. M., Rad. Research Soc. Annual Meeting. Cleveland, Ohio, May 17 (1954).
13. SAFRONOV, A. P., *Zh. Analit. Khim.*, No. 3, 360 (1958).

Effect of Complexing Agents
on the Character of the Binding of Radioisotopes
in the Blood

L. M. Razbitnaya and V. S. Balabukha

Numerous investigations have been devoted to a study of the possibility of removing radioactive isotopes from the body by means of complexing agents. From this point of view a group of diaminopolycarboxylic acids known as complexones attracts special attention.

Experimental investigations show a considerable increase in the elimination from the body both of stable and of radioactive isotopes when these compounds are used. Thus, the value of Na_2 Ca-EDTA was noted, for example, in the therapy of acute lead poisoning[8, 9]. Na_2 Ca-EDTA may be used in poisoning by mercury, copper, cobalt and nickel[1]. Foreman [5, 6] and Cohn, Gong and Fishler[2] used Na_2 Ca-EDTA to accelerate the elimination of [91]Y from the body.

Greenberg[7], in studying the effects of complex-forming agents on the metabolism of stable and radioactive isotopes, found that the distribubution and elimination of [140]La and ^{90}Y, when introduced into the animal in the form of chloride salts, differs markedly from the behaviour of complexes of these elements with EDTA.

D. I. Semenov and I. P. Tregubenko[10] also showed the effectiveness of certain complexones in the removal of heavy and rare earth elements from the body.

G. Ye. Fradkin and V. F. Ushakova (see p. 147) compared the amounts of [91]Y deposited in the organs when the isotope was given in the form of the chloride with those following its administration in the form of complexes with ethylenediaminetetraacetic and cyclo-hexanediaminetetraacetic acids. Less isotope accumulated when a complexed form was given. All the investigators mentioned note the substantial advantage gained by early administration of complex-forming agents, while a large part of the radioisotope is still in the blood, compared with the effect of using them at a later period, i.e. after the radioisotopes have penetrated into the tissues, particularly bone tissue.

For an explanation of the effectiveness of various complexing agents in biological experiments the concept of a change in the character of the binding of isotopes in the blood and bone tissue during the action of complexing agents is of substantial importance.

It should be borne in mind that the same element, when combined with different complexing agents, may behave in different ways in the body, being deposited in different organs and eliminated at different rates.

The present investigation is concerned with the state of the radio-isotopes of yttrium and cerium in the blood during the action of the complexones ethylenediaminetetraacetic acid (EDTA), cyclohexanedi-aminetetraacetic acid (CDTA) and cyclopentanediaminetetraacetic acid (CPDTA).

The elements to be studied, which belong to the lanthanoid group, are characterized on the one hand by a strongly pronounced tendency to complex formation, and on the other by the property of being in a colloidal state at the pH of the blood. These features must be taken into consideration when using any particular complexing agents as means for accelerating the elimination of radioactive isotopes from the body. The lanthanoids, as is known, are divided into three groups: light — La, Sm, intermediate group — Eu, Gd, Tb, heavy — Dy, Lu.

Correspondingly their metabolic behaviour in the body is somewhat different. The first group, i.e. the light lanthanoids, is characterized by a larger deposition in the liver and a smaller one in the skeleton. The intermediate group is deposited approximately equally in the liver and the skeleton. The heavy lanthanoids are deposited mainly in the skeleton[4].

On a basis of metabolic behaviour yttrium may be included in the group of heavy, and cerium in the group of light, lanthanoids.

In order to study the action of complex-forming substances on radioisotopes found in the blood, the electrophoresis method was used, which enabled a determination to be made of what proportion of the radioisotope in the chosen conditions exists in the blood in the form of complex anions (complexes), and what proportion in the form of cations. This method and the apparatus used have been described in detail earlier in our symposium, p. 125.

The experiments were carried out with blood serum which was obtained from rabbits after they had been injected with the radioactive isotopes of yttrium or cerium at 0·1 mc/kg weight. Blood was taken from the aural vein 0·5–1 hr after injection of the isotope, i.e. while the bulk of it was still in the blood.

In the first series of experiments solutions of one of the complexing agents indicated above were introduced *in vitro* into the blood serum 20–30 min before the experiment, in an amount corresponding to the maximum dose of this substance tolerated by the animal: Na salts of CPDTA — 0·5 mg; Ca complexes of EDTA and CDTA — 5 mg/1 ml. serum.

In the second series of experiments the complexing agents were injected intravenously, 30–40 min after injection of the radioactive isotopes, at a dose-level of 250 mg/kg for Ca complexes. Blood was taken from the animals 30–50 min after injection of the complexing agents.

In both series of experiments the blood serum was placed in the central part of the apparatus, and the side limbs were filled with a solution of NaCl (concentration 0·3 per cent, pH 7·4). A direct current of 5–10 mA was passed through the apparatus. That part of the radio-isotope which is combined into a complex anion in the blood, migrates on electrophoresis towards the anode and accumulates in the anode field. The radioisotope in the form of a cation collects in the cathode solution.

At the end of the experiment the liquid from the anode and cathode vessels is transferred to test tubes with ground glass stoppers. Aliquot portions were carefully evaporated on stainless steel planchets and counted.

The results of the first series of experiments are given in Tables 1 and 2. These show the changes in the character of the binding of the isotopes in the blood serum on the introduction of complexing agents.

Studies on the state of ^{91}Y in the blood of rabbits, when given in the form of chloride, have been carried out earlier (see present symposium page 125).

It is seen from the data in Tables 1 and 2 that the complexing agents exert a considerable influence on the state of the radioisotopes of the rare earth elements (yttrium and cerium) in the blood, incorporating them into soluble diffusible compounds. The extent of the incorporation depends largely on the concentration of complexing agent used. Thus,

F

TABLE 1

EFFECT OF COMPLEXING AGENTS, INTRODUCED *in vitro*, ON THE CHARACTER OF
THE BINDING OF ^{91}Y IN THE BLOOD SERUM OF ANIMALS INJECTED
WITH THE RADIOISOTOPE

Complexing agent (acid)	Radioisotope content of solution, % of total		
	In cathode	In central section	In anode
Ethylenediaminetetraacetic (Na, Ca-salt)	6·6	Traces	93·4
Cyclohexanediaminetetraacetic (Na, Ca-salt)	8·4	Traces	91·6
Cyclopentanediaminetetraacetic (Na-salt)	57·5	Traces	42·5

Note — The values in Tables 1–3 are the average of the results of 2–3 experiments. The variation does not exceed 5 per cent.

TABLE 2

EFFECT OF COMPLEXING AGENTS, INTRODUCED *in vitro*, ON THE CHARACTER OF THE
BINDING OF ^{144}Ce IN THE BLOOD SERUM OF ANIMALS INJECTED WITH THE
RADIOISOTOPE

Complexing agent (acid)	Radioisotope content of solution % of total		
	In cathode	In central section	In anode
Ethylenediaminetetraacetic (Na, Ca-salt)	10·3	Traces	89·7
Cyclohexanediaminetetraacetic (Na, Ca-salt)	9·5	Traces	90·5
Cyclopentanediaminetetraacetic (Na-salt)	37·4	Traces	62·6

TABLE 3

ELECTROPHORESIS OF BLOOD SERUM OF ANIMALS INJECTED WITH RADIOISOTOPE
AND COMPLEXING AGENT

(Duration of electrophoresis 70–80 hr)

Complexing agent (acid)	Radioisotope content of solution, % of total		
	In cathode	In central section	In anode
Ethylenediaminetetraacetic (Na, Ca-salt)	2·7	Traces	97·3
Cyclohexanediaminetetraacetic (Na, Ca-salt)	8·2	Traces	91·8

Ca-EDTA and Ca-CDTA, introduced at 5 mg/ml., combine with about 90 per cent of the ^{91}Y. Na-CPDTA, which is equal to CDTA in complex-forming capacity, but injected at 0·5 mg/ml., combines with only about 40 per cent. The results agree with those of Dudley[3] who, when studying the effect of chelating agents on the metabolism of ^{91}Y, found that in the presence of proteins and certain complexing agents (citrate, EDTA, N(2-hydroxyethyl)diaminoethanetriacetic acid, N,N-di(hydroxyethyl)glycine, the radioisotope is incorporated into complex compounds which are dialysed through Cellophane membranes.

In Table 3 the results of the second series of experiments are given, in which the complexing agents were injected intravenously 0·5–1 hr after injection of the radioisotope.

They indicate the state of ^{91}Y in the blood after the injection of complex-formers.

It is seen from the results in Table 3 that the complexones form stable compounds with ^{91}Y and consequently can be used for accelerating the elimination of this element from the body in the early stages after injection. The complexes of ^{144}Ce with ethylenediamine- and cyclohexanediaminetetraacetic acids should be regarded as less stable than those of ^{91}Y. In the first stages of electrophoresis (12 hr), ^{144}Ce, like ^{91}Y, is detected mainly in the form of a complex anion (90 per cent in the anode solution and 10 per cent in the cathode).

When the pH falls, which occurs on more prolonged electrophoresis, the bulk of the ^{144}Ce gradually accumulates in the cathode solution. A similar phenomenon was not observed in the case of ^{91}Y. This is in accordance with the values of log K_i (instability constant) for the elements indicated with the complexing agents tested.

From a study of the condition of the radioisotopes of yttrium and cerium in the blood under the influence of the complexones ethylene-, cyclohexane- and cyclopentanediamine-tetraacetic acids, it was found that ^{91}Y occurring in the blood forms stable soluble complex compounds with the complexones used. ^{144}Ce also gives complex compounds with the complexing agents tested, but they are less stable, decomposing when the acidity is increased.

REFERENCES

1. BERSIN, FH., *Schweiz. Med. Wochensohr.*, **26**, 608 (1953).
2. COHN, S. H., GONG, J. K. and FISHLER, M. C., *Nucleonics*, **11**, 1, 56 (1953).
3. DUDLEY, H. C., *J. Lab. Clin. Med.*, **45**, 5, 793 (1955).
4. DURBIN, P. W., WILLIAMS, M. H., GEE, U., NEWMAN, R. H. and HAMILTON, F. G., *Proc. Soc. Exptl. Biol. Med.*, **91**, 1, 78 (1956).
5. FOREMAN, H., *Arch. Industr. Hyg. Occupat. Med.*, **7**, 137 (1953).
6. FOREMAN, H., *J. Amer. Pharm. Assoc. Sci.*, **52**, 629 (1953).
7. GREENBERG, J., *J. Clin. Invest.*, **33**, 939 (1954).
8. RUBIN, M., GIGNAC, S., BESSMAN, S. P. and BELKNAP, E. L., *Science*, **117**, 3050, 659 (1953).
9. SIDBURY, J. B., BYNUM, J. C. and FETZ, L. L., *Proc. Soc. Exptl. Biol. Med.*, **82**, 2, 226 (1953).
10. SEMENOV, D. T. and TREGUBENKA, I. P., In the symposium *Complexones: synthesis, properties, application in biology and medicine* (Koupleksony: sintez svoistva, primeniye v biologii i meditsine), p. 89, Sverdlovsk, Ural'skii Filial Akad. Nauk, SSSR (1958).

Character and Stability of the Binding of [91]Y in Bone Tissue

N. O. RAZUMOVSKII, O. L. TORCHINSKAYA and
V. S. BALABUKHA

THE QUESTION of the possibility of removing radioactive isotopes of rare earth elements from the body, particularly yttrium, which accumulates in considerable amount in the skeleton, may be decided when there is an adequate knowledge of the character of its binding with the bone tissue.

When [91]Y is administered without a "weighable" quantity of carrier its maximum accumulation in the bones (about 45 per cent) occurs 6 hr after the intravenous injection. By this time the radioactive isotope has practically all disappeared from the blood[10]. The results of Rayner, Tutt and Vaughan[8] also indicate rapid deposition of radioactive yttrium in the skeleton when it is introduced, carrier-free, intravenously. The authors point out that radioactive yttrium combines with bone more stably and is excreted less than radioactive strontium. According to the observations of Hamilton[6], when radioactive yttrium is injected intraperitoneally, up to 50 per cent of the dose introduced is deposited in the skeleton and it is excreted extremely slowly (0·05 per cent per day).

The level of this radioactive isotope in the skeleton remains more or less constant, whereas the radioactivity of the soft tissues falls considerably with time.

It follows from this that the natural excretion of yttrium takes place mainly through a reduction of the level in the soft tissues (liver, spleen, kidneys and so on).

As radioautographs of sections of the bones of rats show, the radioactive yttrium is located in the organic base of the bone in the zone of active growth, under the epiphyseal cartilage at the extremity of the bone[3]. The idea has been put forward that the isotope [91]Y is combined with the mucopolysaccharides[11], but there are no direct experimental results in the literature confirming this hypothesis. This was the reason for carrying out the present work.

141

Rabbits weighing 2·5–3·0 kg were used. ^{91}Y in the form of chloride, without carrier, was injected all at once intraperitoneally at 0·1 mc/kg weight. Animals were decapitated on the 1st, 5th, 20th and 30th days after injection. Femurs freed from bone marrow and adjoining tissues were used for the investigation and were treated by the method of Neuberger and Slack[7]. The method consists in boiling the bone for a short time in water, soaking for a day in a solution of urea, and extracting the fat. The bone is then subjected to enzymatic hydrolysis with trypsin (pancreatin). The collagen remains in the undigested fraction of the bone and products of the enzymatic hydrolysis of other proteins pass into solution.

The radioactivity was assayed in the water-soluble, urea, and lipoid (alcohol-ether soluble) fractions, the fraction hydrolysed by trypsin, the collagen fraction (bone residue not hydrolysed by trypsin) and finally the whole bone.

The results of these experiments are given in Table 1*.

TABLE 1

AMOUNTS OF RADIOACTIVE ISOTOPE ^{91}Y IN DIFFERENT FRACTIONS OF BONE

Time after injection of ^{91}Y, days	Number of animals	Fraction, per cent, of ^{91}Y content of whole bone				
		Water-soluble	Urea	Lipoid	Hydrolysable by trypsin	Collagenic (non-hydrolysable)
1	6	2·88±0·36	2·61±0·74	0·30±0·03	1·14±0·21	94·47±1·45
5	8	2·25±0·64	1·26±0·25	0·32±0·07	0·92±0·13	90·50±1·38
20	6	1·79±0·56	0·61±0·23	0·10±0·02	0·29±0·17	95·0 ±0·97
30	7	1·11±0·28	0	0·24±0·02	0·49±0·19	96·61±0·67

* The values given in Tables 1 and 2 are obtained by statistical treatment of the results from the formula—

$$M \pm \frac{\sigma}{\sqrt{N}}$$

where M is the arithmetic mean,
σ is the standard deviation,
N is the number of observations.

It is seen from the table that less than 3 per cent of the ^{91}Y passes into the water-soluble fraction. The transfer of radioisotope to this fraction decreases with increase of the interval of time after administration.

Thus, if the amount of ^{91}Y in the water-soluble fraction one day after the injection is 2·88 per cent, on the 30th day it is 1 per cent. A similar phenomenon was noted for the urea fraction. The smallest amount of ^{91}Y is contained in the lipoid fraction. In the hydrolysate after enzymatic cleavage of the bone proteins by trypsin there appears from 0·3 to 1 per cent of the radioisotope. The main bulk, however, (90–96 per cent) remains in the collagen fraction not hydrolysed by trypsin.

It was necessary to explain the stability of the binding of ^{91}Y both in the whole bone and in its collagen fraction. For this, samples of pulverized whole bone and collagen were treated with acetate buffer of different pH values, and with hydrochloric acid. The acetate buffer was added at a rate of 15 ml./1 g tissue, and shaken mechanically for 2 hr. After 3 or 4 extractions at the same pH, no radioactivity was found in the last portion of acetate buffer. This indicated that extraction of radioactive yttrium at the given pH was complete. Whole bone was treated with 0·1 N hydrochloric acid (pH 1·3) under similar conditions. Hydrochloric acid of this concentration was also used at 4°C, for 18 hr, on the samples being studied. No appreciable differences were noted in the amounts of radioactivity in the 0·1 N hydrochloric acid solutions. More concentrated hydrochloric acid, 1 N (3·65 per cent), which is generally used for full demineralization of bone, was also used[9]. Demineralization was carried out for 18 hr at 4°C. The results of these experiments are given in Table 2.

The values contained in the table are averages of results obtained after all four periods, as the amounts of ^{91}Y extractable from whole bone, and the collagen fraction, showed no apparent change with time.

The data in Table 2 show that after treatment of the bone tissue and collagen with acetate buffer, a negligible amount of radioactivity is detected in the solution. When the pH of the acetate buffer is reduced to 3·7–2·8, altogether about 4 per cent of the radioactive yttrium can be separated from whole bone and about 1·5 per cent of that in the collagen fraction. The higher figure for the separation of ^{91}Y from whole bone can, apparently, be explained on the ground that in the separation of the collagen fraction the portion of radioactive isotope not bound to it was removed. When $0 \cdot 1$ N hydrochloric acid acts on whole bone, on an average 15 per cent of the ^{91}Y passes into solution. Practically all the radioactive yttrium (about 97 per cent) is separated from the bone only by the action of 3·65 per cent hydrochloric acid. Thus, the results of the experiments indicate the greater stability of the attachment of this isotope to the bone tissue.

TABLE 2

SEPARATION OF ^{91}Y FROM BONE TISSUE AND COLLAGEN FRACTION OF BONE, %

Material	pH of acetate buffer					Hydrochloric acid	
	8	5·4	4·5	3·7	2·8	0·37% (0·1 N)	3·65% (1 N)
Whole bone	—	1·02±0·25	1·39±0·32	3·43±0·24	4·28±0·82	15·12±1·31	96·9±0·80
Collagen fraction	0·77±0·03	0·89±0·07	0·81±0·08	1·36±0·69	1·41±0·40	—	—

In order to study the possibility of extracting ^{91}Y from bone tissue by the direct action of solutions of complexing agents (disodium calcium salts of ethylenediaminetetraacetic (Na$_2$ Ca-EDTA) and cyclo-hexanediaminetetraacetic (Na$_2$ Ca-CDTA) acids), experiments were carried out using chromatographic columns.

Defatted bones, bones from which the fat had not been extracted, crushed bones, uncrushed bones and bone powder were taken for the investigation. The concentration of the solutions of salts of the complexing agents (3·75 g/l.) was based on a calculation of the probable level of complexing agent in the blood when the salt was injected into the animal in a dose of 250 mg/kg weight[2]. The bone tissue was washed for 2–3 days in a chromatographic column with 2 l. of a solution of complexing agent (pH 7·33). Under the optimum conditions, a maximum of 5–6 per cent of the radioisotope passed into solution. The degree of subdivision of the bone, like the preliminary extraction of fat, did not affect the extent of the separation of radioactive yttrium from the bone. A three-fold increase in concentration of the solutions of the complexing agents (11·25 g/l.) enabled altogether 21–28 per cent of the isotope to be washed from the bone. The results of these experiments explain the poor effectiveness of Na$_2$ Ca-EDTA and Na$_2$ Ca-CDTA when they are used in conditions where the yttrium is already fixed in the skeleton. It should be borne in mind also that when complexing agents are injected into animals an effective concentration persists for 1–2 hr.

It was noted above that rare earth elements, particularly yttrium, in bone are possibly combined with the mucopolysaccharides[11]. It was therefore necessary to check whether the collagen fraction which we studied was completely free from osseomucoid and whether small amounts of mucopolysaccharides remained in the unhydrolysed part after hydrolysis with trypsin.

For this purpose, samples of the collagen fraction were analysed for their content of hexosamine. The analyses, carried out by L. T. Tutochkina, gave a negative result, indicating the absence of muco-polysaccharides in this fraction. Thus the hypothesis of the combination of ^{91}Y with mucopolysaccharides was not confirmed by direct experiment.

The reason for the binding of radioactive yttrium with the collagen fraction should probably be sought in the chemical and physico-chemical features of the protein. In contrast to other proteins collagen has in its composition a large amount of proline (14 per cent) and, what is particularly important, hydroxyproline (14 per cent)[4]. The binding of yttrium by collagen, however, is evidently not dependent on the amino-acid composition. It is known that in amino-acid composition the collagen of bone is similar to the collagen of skin[5], and the skin is richer in collagen than bone. In air-dried solid bone there is 18 per cent collagen[4], and in skin about 73 per cent[1]. In addition, collagen is found in other connective tissues. Nevertheless there is no information about the fixation of yttrium by these tissues.

How then is it possible to explain the concentration of yttrium in the collagen structures of bones? In contrast to the collagen of other tissues bone collagen is combined with hydroxyapatite structures. It might be suggested that this complex is capable of combining the yttrium ions by means of the functional groups of collagen and the phosphate groups of hydroxyapatite. On the other hand, the compact structure of the collagen fibrils of bone differs from the porous structure of the collagen of other connective tissues. Such dense packing of the primary fibrils of bone collagen makes it possible for the yttrium to be bound by the functional groups of neighbouring fibrils.

The fact that in a weakly acid medium (pH 2·8) the isotope is split off to a negligible extent argues against the adsorption of yttrium on collagen. Even during treatment with 0·1 N HCl (pH 1·3), when the possibility of the formation of a radiocolloid which might be adsorbed on collagen is completely excluded, the main bulk of the yttrium remains in the collagen fraction (not more than 15 per cent is removed). Only on complete demineralization with 1 N hydrochloric acid, with breakdown of the inorganic structures, does practically all the yttrium pass into solution.

Thus, the investigations showed that the ^{91}Y deposited in the skeleton is found mainly (more than 90 per cent) in the collagen fraction of the bone and is bound firmly to the bone tissue.

In experiments *in vitro* in weakly acid medium (pH 2·8) 4·3 per cent of the ^{91}Y is separated from bone, at pH 1·3 it is 15 per cent, and only on complete demineralization of the bone with 1 N hydrochloric acid is practically all the yttrium liberated.

When complexing agents (Ca-EDTA and Ca-CDTA) at a concentration of 3·75 g/l. act directly on bone, in chromatographic columns, 5–6 per cent of the ^{91}Y passes into solution. If the concentration of the complexing agents is increased three-fold (11·25 g/l.), a maximum of 21–28 per cent of the radioactive yttrium is washed out of the bone.

REFERENCES

1. OREKHOVICH, V. N., *Procollagens, their chemical composition, properties and biological role* (Prokollageny, ikh khimicheskii sostav, svoistva i biologicheskaya rol'), Akad. Nauk SSSR, Moscow (1952).
2. COHN, S. H., GONG, J. I. K. and FISHLER, M. C., *Nucleonics*, **11**, 56 (1953).
3. COPP, D. H., AXELROD, J. and HAMILTON, J. G., *Amer. J. Roent. Rad. Therapy*, **58**, 1, 10 (1947).
4. EASTOE, J. I. E. and EASTOE, B., *Biochem. J.*, **57**, 453 (1954).
5. EASTOE, J. I. E., *The Biochemistry and Physiology of Bone*, p. 81, Academic Press, New York (1956).
6. HAMILTON, J. G., *Radiology*, **49**, 3, 325 (1947).
7. NEUBERGER, A. and SLACK, H. G. B., *Biochem. J.*, **53**, 1, 47 (1953).
8. RAYNER, B., TUTT, M. and VAUGHAN, L., *Brit. J. Exptl. Pathol.*, **34**, 138 (1953).
9. ROGERS, H. I., *Biochem. J.*, **49**, 1, 10 (1951).
10. SCHUBERT, I., FINKEL, M. P., WHITE, M. R. and HIRSCH, G. M., *J. Biol. Chem.*, **182**, 635 (1950).
11. VAUGHAN, I. M., *The Biochemistry and Physiology of Bone*, p. 729, Academic Press, New York (1956).

Analysis of the Effectiveness of Complexing Agents which Accelerate the Elimination of Radioactive Isotopes from the Body

G. Ye. Fradkin and V. F. Ushakova

During the last few years a number of research workers have shown[1-4] that complexing agents, ethylenediaminetetraacetic acid in particular, can accelerate the elimination of certain radioactive isotopes from the body when given soon after the introduction of the radioisotope and less effectively, when given after a longer interval.

The task of the present communication is an analysis of the effectiveness of a number of complexing agents as agents for accelerating the elimination of certain radioactive isotopes from the animal body.

The fundamental property of complexing agents on which their effectiveness depends is their capacity for forming stable compounds with the radioactive isotopes. The stability of the complexes may be studied *in vitro* by physico-chemical methods (potentiometric, polarographic, chromatographic, etc.) and quantitatively expressed by means of the instability constant.

$$K_i = \frac{[Me^{n+}]\,[A^{m-}]}{[MeA^{(m-n)-}]}, \text{ where } m-n<0$$

The equation shows that the instability constant, or dissociation constant, is the ratio of the product of the concentrations of the components of the complex ion to the concentration of the whole complex ion. The lower the value of the instability constant the more stable the complex must be.

A whole series of factors may influence the stability of the complex compounds in the body: the presence in the tissues of various ions, especially phosphate anions, which are capable of splitting weak complex compounds, the pH of the medium, the time for which

147

effective concentrations of the complexing agents remain in the blood and tissues, the presence of enzymes which attack the molecules of complexing agents, and so on. Therefore for judging the possible effectiveness of the complexing agents, an estimate of the stability of the complexes with radioisotopes *in vivo* becomes of great significance, and for this experiments in animals were required.

An indication of the effectiveness of the complexing agents was the extent to which radioisotope was retained in the body after administration in the form of complex, relative to that following administration in the form of a simple salt. The ratio between the isotope content of the main storage organs in the control animals, which received the isotope as a simple salt, and that of the corresponding organs in the experimental animals, which received the isotope as a complex, can be used as an index of the stability of the complex.

The greater the stability of the complex, the smaller must be the amount of isotope retained in the organs of the experimental animals, in comparison with the controls.

The complexes to be studied were prepared by mixing solutions of the complexing agents with solutions of salts of the radioactive isotopes (^{91}Y and ^{239}Pu), and the pH was brought to neutral. In order to create the best conditions for the formation of the complex, the complexones were added in excess. The excess added was always a non-toxic dose of the complexing agent.

The complexes thus prepared *in vitro* were administered to the experimental animals (white rats, males weighing 180–220 g) intraperitoneally or orally. The control animals received the same amount of isotope as an ordinary salt, stabilized under conditions of neutral pH by means of mannitol. At certain times after administration of the radioactive isotopes the animals were decapitated and the isotope content of the experimental and control rats' main storage organs (femur, liver, kidney) was assayed. From these results, expressed as percentages (the controls are taken as 100 per cent), an idea of the effectiveness of the complexing agents was obtained.

METHOD FOR DETERMINING THE
RADIOACTIVITY OF TISSUES

Radioactive yttrium (^{91}Y)

The carcasses of the experimental and control animals were first carbonized, after which they were reduced completely to ashes in a muffle furnace at a temperature of 400°C. The ash obtained was dissolved in 2 ml. concentrated hydrochloric acid, the solution was neutralized by the addition of 1·4 ml. 30 per cent NaOH and made up to 25 ml. with 2 per cent sodium citrate.

Aliquots were transferred to standard aluminium planchets and the radioactivity of the samples was determined on a counting apparatus.

To study the bone tissue the femur was dissected from the body of the animal, cleaned from soft tissue, dried in acetone and reduced to ashes in a muffle furnace at 400°C. The ash was treated by the method described above.

The liver and kidneys were weighed, carefully ground, and samples of 50 mg were distributed evenly on planchets.

Plutonium (^{239}Pu)

Preparation of samples from organs and carcasses was done by S. I. Popov's method. The bones and carcasses were ashed (to a white ash) in a muffle furnace at 400°C. The ash was dissolved in a definite volume of 2 N hydrochloric acid. The soft tissues (liver, kidney) were treated with concentrated nitric acid with subsequent evaporation on a water bath and addition to a given volume of dilute hydrochloric acid. The radioactivity of aliquot samples (0·05 ml.) was assayed on a scintillation counter. Duplicate samples gave good agreement.

RESULTS OF THE INVESTIGATIONS

The stability *in vivo* of the complexes of the radioactive isotope of yttrium (^{91}Y) with nitrilotriacetic (NTA), ethylenediaminetetraacetic (EDTA) and cyclohexanediaminetetraacetic (CDTA) acids was studied.

These complexing agents belong to the class of polyaminocarboxylic acids known as complexones, and they do not normally occur in the body. The chemical formulae of these compounds are given below.

Nitrilotriacetic acid   Ethylenediaminetetraacetic acid   Cyclohexane-diaminetetra-acetic acid

The results given in Tables 1 and 1a indicate the stability of the complexes of NTA, EDTA and CDTA with yttrium.

Owing to the insolubility of the calcium salt of NTA, the sodium salt (20 mg.) was used to prepare the complex with yttrium. EDTA and CDTA were used both as sodium salts (20 mg) and as disodium calcium complexes (50 mg per rat)*.

These quantities of the complexones are an excess, relative to the amount of isotope (5 μc).

Five microcuries of yttrium (^{91}Y) chloride, carrier-free, were given intraperitoneally to each of the control rats, and the same amount of radioactive isotope, in the form of complex, to the experimental rats. The animals were killed on the 10th day after injection.

It follows from the results given in Table 1 that the complexes of yttrium with CDTA are the most stable, then those with EDTA, and the complex of yttrium with NTA is unstable. In the last case the isotope accumulates in the bone tissue as much as, or even somewhat more than, in the control. On comparing the stability of the complexes *in vivo* with the values of the instability constants of NTA, EDTA and CDTA with yttrium *in vitro* (see Table 1a), it is apparent that there is an inverse relationship between the instability constants of the complex compounds and the stability of the complex compounds in the body.

---

* The toxicity of these preparations was tested in white mice. The sodium salts of the complexones are toxic on account of their capacity for causing an acute hypocalcaemia, and a dose of 20 mg is the maximum permissible in a rat. The sodium–calcium salts of the complexones are non-toxic, and a dose of 50 mg in a rat is considerably less than the maximum permissible dose.

The subsequent (2nd) experiment was set up with the object of determining the stability of the complexes of yttrium with NTA, EDTA and CDTA when they were given orally to rats.

TABLE 1

RESULTS OF A STUDY OF THE STABILITY OF COMPLEXES OF ^{91}Y WITH CYCLOHEXANEDIAMINETETRAACETIC, ETHYLENEDIAMINETETRAACETIC AND NITRILOTRIACETIC ACIDS

Organ	^{91}Y Content, % of control				
	^{91}Y+ Na$_2$Ca CDTA	^{91}Y+ Na$_2$Ca EDTA	^{91}Y+ Na CDTA	^{91}Y+ Na EDTA	^{91}Y+ Na NTA
Femur	15·7±0·6	19·0±1·6	18·0±1·0	28·0±4·0	100 ±6·0
Liver	10·0±1·0	12·0±2·0	9·0±1·0	7·0±0·4	16·0±0
Kidney	8·0±1·5	16·0±0·5	6·2±0·3	7·1±0·1	12·0±1·5
Carcass (remainder)	6·9±0·5	10·7±0·1	—	—	—

Note—The values are the averages for 4 rats. In each column of the table the value of the radioactivity of the whole organ is given. The animals were killed on the 11th day after the injection of the isotope.

TABLE 1A

INSTABILITY CONSTANTS OF COMPLEXES OF ^{91}Y WITH CDTA, EDTA AND NTA

Complex	Value of instability constant (K_1)
CDTA + ^{91}Y	$\sim 10^{-23}$
EDTA + ^{91}Y	$\sim 10^{-18}$
NTA + ^{91}Y	$\sim 10^{-13}$

The complexones were used in the form of sodium salts in the same quantities (20 mg each) as in the previous experiment. The results of the 2nd experiment are given in Table 2.

As seen from Table 2, the amounts of ^{91}Y in the bones and liver of the experimental animals were much higher than the amounts in the corresponding organs of the control animals to which the isotope was given orally in the form of chloride. These results are easily explained.

It is known that the radioactive isotopes of rare earth and heavy elements are poorly absorbed from the intestine into the blood, because their simple salts are hydrolysed in the small intestine with the formation of colloidal hydroxides. The absorption of the isotopes of yttrium, plutonium, polonium and so on amounts to hundredths of

TABLE 2

RESULTS OF A STUDY OF THE STABILITY OF COMPLEXES OF YTTRIUM (^{91}Y)
WITH NITRILOTRIACETIC, ETHYLENEDIAMINETETRAACETIC AND
CYCLOHEXANEDIAMINETETRAACETIC ACIDS, GIVEN ORALLY

Organ	^{91}Y Content, % of control		
	^{91}Y + CDTA $K_1 \sim 10^{-23}$	^{91}Y + EDTA $K_1 \sim 10^{-18}$	^{91}Y + NTA $K_1 \sim 10^{-13}$
Femur	166 ± 0.4	308 ± 11	879 ± 0.2
Liver	254 ± 23	586 ± 9	406 ± 0.5
Kidney	78 ± 0.1	107 ± 4.9	145 ± 10

Note—Each figure is the average for 4 rats. The animals were killed on the 11th day after the introduction of the isotope.

one per cent of the dose introduced orally. When the isotopes of these elements are in the form of the soluble complex compounds there is a sharp increase in the absorption of the isotope from the intestine. This may be the case with all radioactive isotopes the salts of which are easily hydrolysed. Therefore in cases of ingestion and inhalation of the isotopes named, oral use of complexing agents is contraindicated. The object of the experiment under discussion, however, was not to confirm this. It is necessary to pay attention to other more essential facts. If the amounts of isotope are compared in the three groups of animals that had received yttrium orally in the form of the three different complexes, it can be seen that the order of stability of the complexes remains the same as it was with intraperitoneal injection. Consequently, when the complexed form of the isotope is adsorbed from the intestine into the blood, the extent to which the radioactive isotope of yttrium is retained in the storage organs bears a similar relationship to the stability of the complex.

In order to confirm the relationship made apparent in experiments 1 and 2, between the values of the instability constants and the stability of the complex compounds of the radioactive isotopes *in vivo*, supplementary experiments 3 and 4 were carried out.

TABLE 3

RESULTS OF A STUDY OF THE STABILITY OF COMPLEXES OF ^{91}Y WITH ETHYLENEDIAMINETETRA-ACETIC AND AMINOMETHYLPHOSPHINIC ACIDS

Organ	^{91}Y content, % of control	
	EDTA + ^{91}Y	AMP + ^{91}Y
Femur	8 ± 0.04	50 ± 26
Liver	4 ± 1.20	70 ± 30
Kidney	20 ± 0.15	70 ± 38

Note — The animals were killed 10 days after the injection of the isotope. Each figure in the table is the average for 4 rats.

In experiment 3 the radioactive isotope of yttrium (^{91}Y) was injected intraperitoneally into white rats in combination with two complexing agents of different chemical types (ethylenediaminetetraacetic and aminomethylphosphinic (AMP) acids). EDTA was given in the form of a calcium salt in a quantity of 50 mg, AMP in the form of a sodium salt in a quantity of 60 mg, which is the maximum amount soluble in 0.5 ml., i.e. in the volume injected into each rat. The relative stability of the complexes formed with yttrium by these complexing agents was previously found *in vitro* by the chromatographic method (L. I. Tikhonova). It was found that *in vitro* the complex of AMP with yttrium was less stable than the complex of the same isotope with EDTA.

The results of the third experiment are given in Table 3.

It follows from the results in Table 3 that the complex of yttrium with EDTA is considerably more stable than the complex with AMP. Such results were obtained in spite of the fact that the molar concentration of AMP exceeded the molar concentration of EDTA by 2·5 times.

When ^{91}Y is injected as a complex with AMP, the larger variation in the results is of biological origin, and is not due to errors in the radioactivity measurements, because different samples gave good agreement. In experiment 4 the radioactive isotope of yttrium was injected intraperitoneally in the form of three complexes, those with

TABLE 4

RESULTS OF A STUDY OF THE COMPARATIVE STABILITY OF COMPLEXES OF ^{91}Y WITH ETHYLENEDIAMINETETRAACETIC, CYCLOHEXENEDIAMINOTETRAACETIC, AND CYCLOHEXANEDIAMINETETRAACETIC ACIDS

Complex	^{91}Y Content, % of control
CHDTA$+^{91}Y$	34 ± 13
EDTA$+^{91}Y$	25 ± 5
CDTA$+^{91}Y$	23 ± 1

Note — Each figure is the average for 4 rats. The animals were killed on the 6th day after the injection of the isotope.

ethylenediaminetetraacetic (EDTA), cyclohexanediaminetetraacetic (CDTA) and cyclohexadienediaminetetraacetic (CHDTA) acids. The last compound is distinguished from CDTA by the presence of double bonds in the ring. All were used in the form of calcium salts at 250 mg/kg, i.e. 50 mg for a rat weighing 180–200 g. The stabilities of the complexes formed by EDTA, CDTA and CHDTA with yttrium were compared *in vitro* by a chromatographic method (L. I. Tikhonova). It was found that CHDTA forms a complex with yttrium which is less stable than the EDTA and CDTA complexes. In Table 4 are given the results of experiment 4 which compares the effectiveness of EDTA, CDTA and CHDTA.

It follows from the results in Table 4 that the complex of CHDTA with yttrium has the least stability *in vivo*, which agrees with the results of the chromatographic investigations *in vitro*.

The experiments showed that of all the complexing agents tested, the most effective is cyclohexanediaminetetraacetic acid (CDTA). Accordingly, attention was centred on certain features of the behaviour of this complexone in the body. In the first place it seemed interesting to find out how long effective concentrations of the calcium salt of CDTA remained in the blood. For this purpose, 25 and 50 mg Na_2 Ca-CDTA were injected intraperitoneally into experimental rats. Then after 0·5 and 1 hr the animals of both groups received 5 μc each

TABLE 5

^{91}Y CONTENT OF THE FEMUR AND CARCASS OF RATS AFTER THE INTRAPERITONEAL INJECTION OF Na_2Ca–CDTA

Organ	^{91}Y content, % of control			
	After 30 min		After 60 min	
	50 mg	25 mg	50 mg	25 mg
Femur	2·0±0	2·0±0	4·0±1·1	8·0±4·0
Carcass (remainder)	2·5±0	—	5·5±0	—

Note — Each figure in the experimental groups is an average for 2 rats, and in the control group for 4.

of yttrium chloride (^{91}YCl$_3$) intraperitoneally. Eleven days later the experimental and control rats were decapitated. The presence of effective concentrations of Na_2 Ca-CDTA in the body was indicated by the ^{91}Y content of the femur and carcass of the experimental animals, this being lower than in the controls.

The results of the experiment are given in Table 5.

It follows from the results in Table 5 that calcium cyclohexanediaminetetraacetate, injected in a dose of 0·25 g/kg and 0·125 g/kg, 0·5 and 1 hr before the radioactive isotope of yttrium (^{91}Y) is given, shows great effectiveness, preventing the deposition of the isotope in the organs and tissues.

In view of these results, it was decided to test the prophylactic effect of Na_2 Ca-CDTA when given orally, against radioactive yttrium

(^{91}Y). For this purpose, an experiment (6) was carried out with two groups of animals.

The first group of experimental rats was divided into two sub-groups. One sub-group received orally 100 mg (1·0 ml. 10 per cent solution) of Ca-CDTA, and the other a similar dose of the analogue, cyclopentanediaminetetraacetic acid (Na$_2$ Ca-CPDTA).

The second group, which was also divided into two sub-groups according to the agent used, received two doses, the second dose being given 4 hr after the first dose.

TABLE 6

CONTENT OF ^{91}Y GIVEN 24 HR AFTER A SINGLE ADMINISTRATION OF Na$_2$Ca–CDTA AND Na$_2$Ca–CPDTA

Organ	^{91}Y content, % of control	
	Ca–CDTA	Ca–CPDTA
Femur	60±12	76±21
Liver	101±1·4	168±30
Kidneys	—	90±0·2

Note—In Tables 6–8 each figure is the average for 4 rats.

Twenty-four hours after the first dose of prophylactic both groups of experimental animals were injected intraperitoneally with 5 μc of yttrium chloride (^{91}YCl$_3$). The control animals simultaneously received the same amount of isotope. The experimental and control animals were killed on the 10th day after the injection of the yttrium and the amount in the main storage organs was determined.

The results of the experiment are given in Tables 6 and 7.

It follows from the results in Tables 6 and 7 that the double oral administration of calcium cyclohexanediaminetetraacetate and cyclo-pentanediaminetetraacetate is more effective than the single administration. The double administration of the complexones decreases the deposition of radioactive yttrium not only in the bones but also in the soft tissues (liver, kidney).

TABLE 7

Content of ^{91}Y Given 24 hr after a Double
Administration of Na$_2$Ca–CDTA and Na$_2$Ca–CPDTA

Organ	^{91}Y content, % of control	
	Ca–CDTA	Ca–CPDTA
Femur	70·6±9	62·6±1·2
Liver	52·0±0·09	57·2±16·0
Kidneys	52·0±0	71·0±0·002

The next experiment, the 8th, was performed in order to study the comparative stability *in vivo* of the complexes of plutonium with ethylenediaminetetraacetic (EDTA) and cyclopentanediaminetetra-acetic acids (CPDTA). To prepare the complexes, 50 mg of each of the agents, in the form of the calcium salt, was used. Plutonium was injected intraperitoneally into the experimental rats in the complexed

TABLE 8

Results of a Study of the Stability of Complexes of
^{239}Pu with Ethylenediaminetetraacetic and
Cyclopentanediaminetetraacetic Acids

Organ	^{239}Pu content, % of control	
	EDTA	CPDTA
Femur	276±0·4	133±11
Liver	28±13	27±3·3
Kidneys	155±0·01	179±8
Carcass (remainder)	128±0·01	64±0·001
Whole rat	76	45·9

Note — The animals were killed on the 5th day after the injection of the isotope.

form, in a dose of 1 μc. The control rats received the same amount of isotope in the form of a simple nitrate stabilized with mannitol. Four days after the injection of the plutonium the experimental and control rats were killed and the amounts of isotope in the storage organs (femur, liver, kidney) and in the rest of the carcass were determined.

The results of this experiment are given in Table 8. Table 8a gives the values of the instability constants *in vitro* of the complexes of plutonium (^{239}Pu) with ethylenediaminetetraacetic, cyclohexanedi-aminetetraacetic and cyclopentanediaminetetraacetic acids.

TABLE 8A

INSTABILITY CONSTANTS OF COMPLEXES (OF ^{239}Pu) WITH CYCLOHEXANEDIAMINETETRAACETIC, CYCLO-PENTANEDIAMINETETRAACETIC AND ETHYLENEDIAMINETETRAACETIC ACIDS

Complex	Value of instability constant
CDTA+Pu	$\sim 10^{-29}$
CPDTA+Pu	$\sim 10^{-29}$
EDTA+Pu	$\sim 10^{-24}$

It follows from the results given in Tables 8 and 8a that when plutonium is given in the form of complexes with ethylenediamine-tetraacetic and cyclopentanediaminetetraacetic acids most of the isotope accumulates in the bones and a much smaller proportion is concentrated in the liver.

The amounts of plutonium retained by the whole body after injection of the two complexes are related to the values of the instability constants of the complexes. The lower the instability constant, the smaller is the isotope content of the whole body.

CONCLUSION

Analysis of the experimental results makes it possible to show the main factors determining the effectiveness of complexing agents as means for accelerating the elimination of certain radioactive isotopes of rare earth (^{91}Y) and heavy (^{239}Pu) elements from the animal body.

The first factor is the degree of stability of the complexes formed with a given isotope. The degree of stability is expressed by the value of the instability constant.

The results of experiments 1–4 and 8 confirm the existence of a relationship between the values of the instability constants and the extent to which the radioactive isotopes ([91]Y and [239]Pu), injected into the body as complexes, are retained in the tissues.

The second factor is the degree of stability of the binding of the radioactive isotopes with the tissues. The complexing agents exert their action in competition with the tissues, particularly with the proteins, in the "capture" of the isotopes. The greater the affinity of the radioactive substances for the biological compounds the smaller will be the effectiveness of the complexones. The necessity for bringing this second factor into account is seen on comparing the data given in Tables 1 and 1a on the one hand with those in Tables 8 and 8a on the other. These data (see Tables 1a and 8a) indicate that with ethylenediaminetetraacetic and cyclohexanediaminetetraacetic acids plutonium forms complexes which are 6 orders, i.e. 1 million times, more stable than the complexes formed by the same complexones with yttrium.

From the values of the instability constants it would have been expected that on injection of plutonium and yttrium into the body as complexes with EDTA the amount of [239]Pu retained in the tissues would be much smaller than that of [91]Y. In reality, however, another picture was obtained.

When yttrium was injected as the complex with EDTA 6·2 per cent of the dose was retained (see Fig. 1), whereas on injection of plutonium in the form of the complex with the same complexone 52 per cent of the dose was retained (Fig. 2).

Such results can only be explained on the ground that plutonium is more strongly bound with the tissues than yttrium.

Complexes with radioactive isotopes, however stable, dissociate to some extent, and therefore in the body there will always occur reversible reactions in which the tissues and the complexones compete for capture of the isotope. These reactions may be represented by the following equations:

$$[MeA] \rightleftarrows A + Me + B \rightleftarrows [MeB]; \quad (1)$$
$$[MeA] \leftarrow A + Me + B \rightleftarrows [MeB]; \quad (2)$$
$$[MeA] \rightleftarrows A + Me + B \rightarrow [MeB]; \quad (3)$$

where A is the complexing agent; B is the tissue; Me is the isotope; MeA is the complex ion; MeB is the compound of the isotope with the tissues.

Equation (1) represents the equilibrium existing with certain concentrations of A and B. Equation (2) represents those cases where the effective concentration of the complexone is sufficient to shift the equilibrium towards incorporation of the isotope into the complex. Equation (3) shows the case in which owing to the pronounced

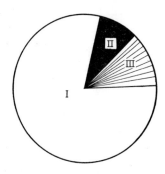

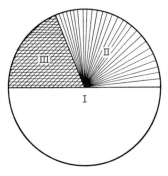

I — control animals, into which the simple salt $^{91}YCl_3$ was injected (content 57 per cent); II — animals into which the yttrium was injected as complex with CDTA (content 4·2 per cent); III — animals into which the yttrium was injected as complex with EDTA (content 6·2 per cent)

Fig. 1.   Radioactive yttrium content of the body, per cent of the dose introduced

I — control animals, into which the plutonium was injected in the form of a simple salt (content 80 per cent); II — animals into which the plutonium was injected as complex with EDTA (content 52 per cent); III — animals into which the plutonium was injected as complex with CPDTA (content 31 per cent)

Fig. 2. ^{239}Pu content of the body, per cent of the dose introduced

affinity of the isotope for the tissues the equilibrium is shifted to the opposite side, i.e. combination of the isotope with the tissue predominates. Hence it becomes clear that the effectiveness of complexing agents cannot be considered without taking into account properties and peculiarities of the isotope in question. In particular, the assessment of the effectiveness of complexing agents in experiments must be different in the cases of yttrium and plutonium. Let us first consider the experimental results with yttrium (^{91}Y). The results of the first experiment (see Fig. 1) showed that if the stability of the complex with radioactive yttrium corresponds to a value for the instability constant of about 10^{-23}, then about 4 per cent of the dose of isotope introduced is retained in the body, whereas in the control about 57 per cent is retained.

When comparing the stability of the complexes of EDTA and CDTA with yttrium *in vivo*, it may be noted that *in vitro* the complex of CDTA with yttrium is 100 thousand times, i.e. five orders, more stable than the complex with EDTA. However, the effect of this is to give only a slight difference in stability *in vivo*, the ratio of the stabilities *in vivo* being about 1·5 : 1, i.e. within the limits of one order.

In view of the above, it may be supposed that the retention of the yttrium isotope in the body would be decreased considerably only when the instability constant of the complex was decreased by another five orders from 10^{-23} to 10^{-28}.

We will consider from the same aspect the results of experiment 8 in which plutonium was injected into the animal in the form of complexes with ethylenediaminetetraacetic and cyclopentanediaminetetra-acetic acids. From the results of the experiment some substantial conclusions may be drawn.

As an indication of the stability of the plutonium complex in the body, corresponding to a value for the instability constant of about 10^{-29}, 31 per cent of the dose of the isotope is retained as a result of injection of the complex (CPDTA $+^{239}$ PU), and in the control 80 per cent.

In vitro the complex of CPDTA with plutonium is 1 million times more stable than the EDTA complex. *In vivo* the stability of the CPDTA complex is about 1·6 times greater than that of the complex with EDTA, i.e. as in the case of yttrium, the difference in stability does not go beyond the limits of one order. It may be considered probable that the retention of the isotope in the body, introduced as a complex, will decrease perceptibly only if the instability constant of the complex decreases by another six orders from 10^{-29} to 10^{-35}.

Of a number of complexing agents tested, the most effective are cyclohexanediaminetetraacetic and cyclopentanediaminetetraacetic acids. These may be used as a reliable first-aid treatment against the acute effects of the radioactive isotopes of yttrium.

Analysis of the experimental results provides a basis for setting the synthetic chemists the clearly defined task of making new complexones.

Very recently there has appeared in print a report[3] on new, very effective complexones which accelerate the removal of the radioactive isotope of cerium (^{144}Ce) from the body. According to the author's data, the disodium calcium salt of 2 : 2′-bis[di(carboxymethyl)amino]-diethyl ether considerably surpasses disodium calcium ethylenedi-aminetetraacetate and cyclohexanediaminetetraacetate in effectiveness.

The task for further investigations must consist in research on and synthesis of new, highly effective complexing agents which, with the radioactive isotopes of rare earth elements, give complexes with instability constants of about 10^{-28} or less, and with isotopes of heavy elements (^{239}Pu), of 10^{-35} or less.

REFERENCES

1. DUDLEY, C., *J. Lab. Clin. Med.*, **45**, 5, 792 (1956).
2. COHN, S. H., GONG, I. K. and FISHLER, M. C., *Nucleonics*, **11**, 1, 56 (1953).
3. CATSCH, A. and DU KHUONG LE, *Nature*, **180**, 4586, 609 (1957).
4. SCHUBERT, I., *Ann. Rev. Nucl. Sci.*, **5** (1955).

Index